Applied Econometrics Association Series

General Editors: **Jean H. P. Paelinck**, Emeritus Professor, Erasmus University, Rotterdam, and **Henri Serbat**, Chamber of Commerce, Paris

The vital importance of econometrics for understanding economic phenomena is increasingly recognised in every field of economics.

The discipline is based on 'scientific processes which aim to identify, explain and forecast economic phenomena using formalised tools to produce, measure, structure and model the information' (Gerard Duru and Henri Serbat, 1991).

The Applied Econometrics Association, established in 1974, seeks to develop the use of econometric tools by producing regular updates on the state of the art and the progress made in each specific field, and so to further the transformation of unconnected facts into pertinent information for the use in analysis and decision-making.

The series was conceived to form a set of working references for advanced students and researches in each specific field, as well as a guide to development across the discipline more broadly.

This exchange of knowledge will be continued permanently by the opening of a debate-site on the Internet [http://www.lyonl.fr/aea].

Titles include:

Françoise Bourdon and Jean Bourdon (editors)
WAGE ECONOMETRICS AND MODELLING

Fabrizio Carlevaro and Jean-Baptiste Lesourt (editors)
MEASURING AND ACCOUNTING ENVIRONMENTAL NUISANCES AND SERVICES

Arthur Getis, Jesús Mur and Henry G. Zoller (editors)
SPATIAL ECNOMETRICS AND SPATIAL STATISTICS

Siv S. Gustafsson and Danièle E. Meulders (editors)
GENDER AND THE LABOUR MARKET
Econometric Evidence of Obstacles to Achieving Gender Equality

Hans Heijke and Joan Muysken (editors)
EDUCATION AND TRAINING IN A KNOWLEDGE-BASED ECONOMY

Sandrine Lardic and Valérie Mignon (editors)
RECENT DEVELOPMENTS ON EXCHANGE RATES

Pavlos Karadeloglou (editor)
EXCHANGE-RATE POLICY IN EUROPE

Emile Quinet and Roger Vickerman (editors)
THE ECONOMETRICS OF MAJOR TRANSPORT INFRASTRUCTURES

Philippe Thalmann and Milad Zarin-Nejadan (editors)
CONSTRUCTION AND REAL ESTATE DYNAMICS

Applied Econometrics Association
Series Standing Order ISBN 0-333-71460-1
(outside North America only)

You can receive future titles in this series as they published by placing a standing order. Please contact your bookseller or, in case of difficulty, write to us at the address below with your name and address, the title of the series and the ISBN quoted above.

Customer Services Department, Macmillan Distribution Ltd, Houndmills, Basingstoke, Hampshire RG21 6XS, England

Recent Developments on Exchange Rates

Edited by

Sandrine Lardic

and

Valérie Mignon
University of Paris
France

palgrave
macmillan

First published 2004 by
PALGRAVE MACMILLAN
Houndmills, Basingstoke, Hampshire RG21 6XS and
175 Fifth Avenue, New York, N. Y. 10010
Companies and representatives throughout the world

PALGRAVE MACMILLAN is the global academic imprint of the Palgrave Macmillan division of St. Martin's Press, LLC and of Palgrave Macmillan Ltd. Macmillan® is a registered trademark in the United States, United Kingdom and other countries. Palgrave is a registered trademark in the European Union and other countries.

ISBN 1–4039–3487–8

This book is printed on paper suitable for recycling and made from fully managed and sustained forest sources.

A catalogue record for this book is available from the British Library.

Library of Congress Cataloging-in-Publication Data
Recent developments on exchange rates / edited by Sandrine Lardic and Valérie Mignon.
p.cm – (Applied Econometrics Association series)
Includes bibliographical references and index.
ISBN 1–4039–3487–8
1. Foreign exchange rates–Mathematical models. I. Lardic, Sandrine. II. Mignon, Valérie. III. Series.

HG3823.R43 2004
332.4'56'015195–dc22

2003067266

10 9 8 7 6 5 4 3 2 1
13 12 11 10 09 08 07 06 05 04

Printed and bound in Great Britain by
Antony Rowe Ltd, Chippenham and Eastbourne

Contents

List of figures

List of tables

Notes on the contributors

Wolfram Berger is a lecturer at the University of Hagen, Germany. He studied Economics at the Universities of Oldenburg and Münster. From 1997 to 2001 he was a junior lecturer at the University of Hagen where he received his Ph.D. from in 2001. His main fields of research are international monetary economics and international finance. Presently, his work is concerned with the international welfare effects of monetary and exchange rate policy as well as with the optimal monetary and exchange rate policy design.

Geir Bjønnes is Associate Professor at the Norwegian School of Management and research fellow at the Stockholm Institute for Financial Research. He obtained his Ph.D. at the Norwegian School of Management, University of Oslo in 2001. He has research interests in market microstructure, Treasury auctions and international finance. In ongoing projects he studies bidder behavior in discriminatory and uniform price Treasury auctions using bidder specific data. In a co-authored project with Dagfinn Rime, he studies dealer behavior in the FX market with data on transactions and dealer inventories. Recently, he has started to study how microstructure data can be exploited to address macro related issues in a co-authored project with Dagfinn Rime and Haakon Solheim.

Hafedh Bouakez is a Senior Analyst at the International Department of the Bank of Canada. He received a Ph.D. in Economics from Université de Montréal (Canada) in 2003. His current research focuses on macroeconomics, international finance, and monetary theory and policy.

Mario Cerrato received his BSc in Economics at the University of Salerno (Italy) and his MSc in Economics at the London Guildhall University. Currently he is Lecturer in Finance at the London Metropolitan University. Previously, he was Visiting Lecturer at the University of Salerno (Italy) and Westminster University (London). He has also been financial consultant at the Banca del Salento (Italy). His current research interests include econometric analysis of dynamic panels, unit root and cointegration, exchange rate modelling, value at risk, and credit derivatives.

Christophe Chamley is Professor of Economics at Boston University, and Directeur de Recherche at EHESS-DELTA (France). He has research interests in social learning and coordination, macroeconomics, and public finance.

Yin-Wong Cheung is Professor of Economics at the University of California, Santa Cruz. He also is a research fellow of the CESifo in Germany. Cheung's areas of research include econometrics, applied econometrics, exchange rate dynamics, asset pricing dynamics, and output fluctuation. He has published over 60 refereed articles in more than 30 professional journals.

Paul De Grauwe (56) is Professor of International Economics at the University of Leuven, Belgium. He was a member of the Belgian parliament from 1991 to 2003. He is doctor honoris cause of the University of Sankt Gallen (Switzerland) and of the University of Turku (Finland). He obtained his Ph.D. from the Johns Hopkins University in 1974. His research interests are international monetary relations, monetary integration, theory and empirical analysis of the foreign-exchange markets, and open-economy macroeconomics. His published books include *The Economics of Monetary Union* and *International Money. Post-war Trends and Theories.*

Ulf Erlandsson is currently a Ph.D. student at the Department of Economics, Lund University. His main research interests are applied time-series analysis, international finance in general, and exchange rate modelling in particular. A substantial part of his research has involved Markov regime switching models with applications to interest- and exchange rates.

Marianna Grimaldi got her undergraduate degree at University of Naples (Italy). Since 1999, she has been working with Paul De Grauwe as Ph.D. student at University of Leuven (Belgium) on models of exchange rate. In particular, her thesis concerns the relationship between the exchange rate and its fundamentals in a nonlinear framework with heterogenous agents. Since few weeks, she has joined the Research Department of the Sveriges Riksbank in Stockholm.

Sandrine Lardic is Senior Lecturer in Economics at the University of Paris X (France) and is a member of the MODEM-CNRS research group. She also is consultant for Sinopia Asset Management. She carries on research concerning econometrics in the fields of financial markets and macroeconomics. She is the author of books and papers on theoretical and applied econometrics.

Valérie Mignon is Professor of Economics at the University of Paris X (France) and is a member of the THEMA-CNRS research group. She conducts research on theoretical econometric methods and applied econometrics in the fields of financial markets and macroeconomics. Her research interests include long-memory processes and nonlinear econometrics. She is the author of books and papers on theoretical and applied econometrics.

Dagfinn Rime is a Researcher at Norges Bank (central bank of Norway) and at Stockholm Institute for Financial Research. He received his Ph.D. from the Norwegian School of Management, and his current research focuses on international finance and on the market microstructure of foreign exchange markets.

Nicholas Sarantis is Professor of International Finance, and Director of the Centre for International Capital Markets, in the Department of Economics,

Finance & International Business at London Metropolitan University, UK. He received a Ph.D. in Economics from the University of Manchester. Past appointments include academic posts, work for an international bank, and consultancy on macroeconomic modelling and forecasting. He has published numerous papers in academic journals on international finance and applied macroeconomics. His current research interests include exchange rate modelling with nonlinear and panel methods, economic and financial forecasting with BVAR models, and the effects of central bank independence on economic performance.

Haakon O. Aa. Solheim is a Ph.D. student at Norwegian School of Management. His research interests include exchange rate economics, monetary economics, and international macroeconomics. He has his master from Norwegian School of Economics and Business Administration and has worked for the Norwegian Ministry of Finance.

Acknowledgments

This book is a selection of articles presented at the LXXXI Applied Econometrics Association international conference on exchange rates held in Marseilles (France), in March 2003.

We would like to thank Céline Baujard and Henri Serbat from the Applied Econometrics Association for helpful assistance. Helpful assistance was also provided by the local organizer, the GREQAM-CNRS research group, which is gratefully acknowledged.

We are also indebted to various members of the scientific committee: Enrique Alberola, Sanvi Avouyi-Dovi, Oscar Bajo-Rubio, Michel Beine, Catherine Bruneau, Zsolt Darvas, Rebecca Driver, Gilles Dufrénot, Christoph Fischer, Eric Girardin, Pavlos Karadeloglou, Iikka Korhonen, Victor Kotlan, Vêlayoudom Marimoutou, Nikolaos Panigirtzoglou, Anne Péguin-Feissolle, Georges Prat, Henri Serbat and Simon Sosvilla-Rivero. We would like to thank Agnès Bénassy-Quéré and Balazs Egert for helpful comments on some papers.

Finally, we would like to thank all participants at this conference, speakers, discutants, and chairman sessions.

Introduction

By Sandrine LARDIC[1] and Valérie MIGNON[2]

Recent developments on international stock markets, together with the uncertain economic environment have led to new approaches in the study of financial markets and, in particular, foreign exchange markets. In this context, the aim of this book is to present recent developments in exchange rate analysis.

The book consists of three parts examining recent developments relating to exchange rates. In the first part, exchange rate models are proposed. In chapter 1, Paul De Grauwe and Marianna Grimaldi develop a model which combines heterogenous agents and transaction costs. Agents have different beliefs about the future exchange rate, which lead them to process information differently. Moreover, due to the presence of transaction costs, the exchange rate dynamics is nonlinear. This model, based on a mean-variance optimization framework, replicates the empirical puzzles and anomalies uncovered in the last decade by the empirical exchange rate literature. More specifically, three main results emerge from this model. First, concerning the disconnect puzzle, it appears that the exchange rate is disconnected from its fundamentals most of the time. This disconnection can be explained by the interactions between heterogenous agents. The second puzzle relates to the fat tails and excess kurtosis which are generated by the model of Paul De Grauwe and Marianna Grimaldi. Indeed, the exchange rate dynamics is characterized by an unpredictable alternation of low and high turbulence periods. Finally, the authors show that chartism does not disappear. This result is consistent with the empirical evidence concerning the importance of chartism in foreign exchange markets. Moreover, chartism may lead to rational forecasting rules since chartists create an informational environment: they create noisy information that becomes the source of profitable speculation.

In chapter 2, an exchange rate model is also proposed by Hafedh Bouakez. However the structure of this model is very different from the previous one. Indeed, the author develop a dynamic general-equilibrium two-country sticky-price model in order to study the real exchange rate dynamics. To generate

[1]MODEM-CNRS, University of Paris X, 200 avenue de la République, 92001 Nanterre Cedex, France, Email: Sandrine.Lardic@u-paris10.fr.

[2]THEMA-CNRS, University of Paris X, 200 avenue de la République, 92001 Nanterre Cedex, France. Email: Valerie.Mignon@u-paris10.fr.

persistent fluctuations in the real exchange rate, a Calvo-type price setting is introduced such that each firm has a constant probability of changing its price in every period. To assess the ability of sticky-price models to generate endogenous persistence phenomena, an analytical approach is adopted. Using such an approach, Hafedh Bouakez shows that real exchange rate persistence is pinned down by the probability of not changing prices. In other words, the rate at which the response of the real exchange rate to a monetary shock dies out, and the rate at which prices adjust are exactly the same. This result indicates that standard sticky-price models do not generate endogenous real exchange rate persistence. This suggests that other factors than price rigidities might affect real exchange rate dynamics. Such factors could be technology and fiscal shocks which are likely to be highly persistent and could account for real exchange rate persistence.

The second part of the book deals with econometric aspects of the exchange rate dynamics linked to regime-switching and structural breaks. In chapter 3, Yin-Wong Cheung and Ulf Erlandsson adopt a nonlinear framework in order to study real exchange rate dynamics. More precisely, the aim of their paper is to investigate whether real exchange rates follow Markov switching dynamics. In this context, the main issue relates to the determination of the number of regimes in the empirical Markov switching model. Since without formal evidence of the presence of multiple regimes, estimation of Markov switching processes may lead to spurious results, Yin-Wong Cheung and Ulf Erlandsson propose a statistical test for the number of regimes. Moreover, due to the fact that this test is a non-standard statistical inference problem, the authors proceed to a Monte Carlo experiment. This test is applied on three real exchange rate series: USD/DM, USD/BP and USD/FF. The period spans from 1973 to 1997 and three frequencies are considered: monthly, quarterly, and annual. While annual and quarterly data yield inconclusive or weak evidence on the presence of Markov switching behavior, monthly data strongly indicate the existence of multiple regimes in real exchange rate data. This conclusion shows that the frequency of observations plays a very important role in revealing regime-switching behavior. Thus, in order to investigate the possibility of Markov switching dynamics, the use of high frequency data is recommended.

The next chapter also takes place in a nonlinear framework. Indeed, Mario Cerrato and Nicholas Sarantis investigate the presence of structural breaks and unit roots in black market real exchange rates from various emerging countries. This is an important topic in the purchasing power parity (PPP) context since the existence of persistent deviations from PPP would question the validity of many theoretical models and have important policy implications. Moreover, as argued by Mario Cerrato and Nicholas Sarantis, it seems important to recall that emerging countries have experienced different exchange rate regimes and important policy shifts and devaluations over the last three decades. These events can cause structural shifts in the real exchange rate underlying process. This highlights that the use of unit root tests to investigate the PPP hypothesis must allow for structural breaks. Using the methodologies of Banerjee, Lumsdaine and Stock (1992)[3] and Zivot and Andrews (1992)[4] on twenty monthly black market exchange rates, the authors find considerable evidence of structural breaks, which are determined endogenously. More especially, real exchange rates are break-trend stationary in nine emerging markets (Bolivia, Ghana, Korea, Malaysia, Nepal, Pakistan, Philippines, Sri Lanka, and Thailand). In most of these countries, the break-trend process is characterized by a shift in the slope and the trend function. For the other eleven countries, the test fails to reject the unit root hypothesis, implying persistent deviations from PPP (Algeria, Colombia, Dominican Republic, Egypt, Ethiopia, Hungary, India, Indonesia, Kenya, Turkey, and Venezuela). While the evidence on PPP from black market exchange rates is mixed, this paper highlights the need to allow for structural breaks in applications of unit root tests to real exchange rates in countries that have experienced significant shifts in exchange rate and in policy regimes.

In the third part of the book, an analysis of speculative attacks and monetary policy is presented. The three chapters which constitute this part are very different. In chapter 5, Geir Bjønnes, Dagfinn Rime and Haakon Solheim propose an analysis of the speculative attacks that affected Norway and Sweden in 1998. More precisely, the authors empirically address two issues in the literature on speculative attacks: the role of foreign speculators and the role of contagion between markets. In order to analyze the impact of an external event, like the Russian moratorium, on small currency markets — the Swedish krona and the Norwegian krone — two unique data sets on actual transactions from Norges Bank and Sveriges Riksbank are used. Three sets of explanatory variables are considered: changes in accumulated net holdings of local currency by foreign financial investors, international risk as measured by the spread between a 10-year swap in the US market and the 10-year yield on US government bonds, and monetary policy reaction through interest rate setting. Various conclusions can be made from this study. First, Geir Bjønnes, Dagfinn Rime and Haakon Solheim find that all three factors influence the exchange rate dynamics. More specifically, the three sets of explanatory variables are found to be highly positively correlated during periods of crisis, suggesting evidence of contagion between markets. Second, the link between international risk and small exchange rates during periods of international uncertainty seems to depend on changes in position of financial investors, and especially foreign financial investors. The last result relates to the role of monetary policy. The authors find that this role is only minor for the outcome of the crisis and the dynamics of the exchange rate does not seem to be very affected by the monetary policy regime.

In chapter 6, Wolfram Berger also deals with topics linked to monetary policy and exchange rate regimes. He addresses the question whether the exchange rate

[3] Banerjee, A., Lumsdaine, R.L. and Stock, J.H. (1992), "Recursive and sequential tests of the unit root and the trend break hypotheses: Theory and international evidence", *Journal of Business and Economic Statistics*, 10, 271-287.

[4] Zivot, E. and Andrews, D.K. (1992), "Further evidence on the great crash, the oil-price shock, and the unit root hypothesis", *Journal of Business and Economic Statistics*, 10, 251-270.

dynamics is consistent with the regimes that governments officially declare they are running. More precisely, starting from the stylized fact that a fear of floating is prevalent in developing countries, Wolfram Berger constructs a stochastic macroeconomic model to investigate why and when countries prefer to limit the flexibility of their officially floating exchange rate. Three key features of this model are a credit constraint for the corporate sector, a currency mismatch in firms' balance sheets and an important pass-through of exchange rate changes to inflation. In this model, with a policymaker seeking to stabilize output and inflation around some target values, a fear of floating is derived as the result of the policymaker's optimal response to real and monetary shocks. Moreover, the author addresses the question whether flexible exchange rates are still effective in stabilizing output if a large portion of debt is dollarized. In this context, the model developed by Wolfram Berger puts forward that foreign currency debt does not render the exchange rate an ineffective tool. Dollarized liabilities reduce the efficiency of flexible exchange rates as a shock absorber. Finally, it is shown that monetary policy is less efficient due to the rise of the costs of the optimal output stabilization.

In the last chapter, Christophe Chamley also adopts a theoretical approach and proposes a canonical model of coordination in which agents act strategically in a multi-period framework with a regime of fixed exchange rates. This model includes a large number of agents, and any speculator in a currency attack wonders how many other agents believe that the fundamentals are such that an attack will succeed, how others think about others' beliefs, and so on. Christophe Chamley assumes that the mass of speculators is an unobservable parameter. Two states of nature are considered: the "high" state in which the mass of speculators is sufficiently large to induce a devaluation if all agents buy the foreign currency, and the "low" state in which this mass is "sub-critical". Thus, one of the main characteristics of the model is the dynamic setting for speculative attacks when agents have different informations. Within this context of coordination problem, there is some incentive to delay in order to get more information on others. But, if an agent delays too much, he comes too late for the gain of the attack and does not get anything. To sum up, this model incorporates a trade-off between "going early" with little information, and delaying for more information with the risk of coming too late. In all periods, there will be an equilibrium with no speculative attacks. Under some conditions on the beliefs of the agents, there will also be other equilibria with speculative attacks. The equilibrium strategy which is analyzed by Christophe Chamley converges to a stationary solution when the number of periods goes to infinity.

Part I

Exchange rate models

Chapter 1

Heterogeneity of beliefs and exchange rate dynamics

By Paul DE GRAUWE[1] and Marianna GRIMALDI[2]

Abstract

In this paper we develop a model of the exchange rate. Agents have different beliefs about the future exchange rate. In addition, the existence of transaction costs introduces an important non-linearity. We show that our model creates complexity in the market which is characterized by the fact that the exchange rate is disconnected from its fundamental most of the time. Periods of tranquility and turbulence alternate in unpredictable manner. Finally, we show why chartists forecasting rules can be rational.

1.1 Introduction

The efficient market paradigm has dominated exchange rate modelling. However the empirical evidence has not been favorable for this paradigm. A number of empirical puzzles and anomalies have been detected casting doubt about the validity of the efficient market hypothesis. As a result, economists have felt the need to search for alternative modelling approaches. Some models have focussed on the heterogeneity of agents, *i.e.* agents hold different beliefs which lead them to process information differently (Kurz (1994), Kurz and Motolese (2000)). Other models take into account the existence of transaction costs in the goods markets which may lead to non linearities in the exchange rate dynamics (Kilian and Taylor (2001), Taylor *et al.* (2001)). Although these two approaches have been growing mostly apart we believe that there is a need to combine these two insights for a better understanding of the exchange rate dynamics.

[1]University of Leuven, Naamsestraat 69, 3000 Leuven, Belgium.
[2]University of Leuven, Belgium.

Recently heterogeneity of agents was also introduced in rational expectations models (see *e.g.* Bacchetta and van Wincoop (2003)). The implication of rational expectations in models with heterogeneous agents is that it creates "infinite regress", *i.e.* the exchange rate depends on the expectations of other agents' expectations, which depends on the expectations of the expectations of other agents' expectations, and so on, ad infinitum. This leads to intractable mathematical problems except under very restrictive simplifying assumptions. Although this approach is intellectually satisfying, it is unclear that it is a good representation of what agents do in the exchange market. It requires these agents to solve a mathematical problem to which mathematicians have as yet been unable to give a general solution. This seems to us as imposing too large an informational burden on individual agents.

Our approach contrasts with this rational expectations approach in that agents use simple rules, the "fitness" of which is then controlled *ex-post* by checking their profitability, and by switching to the more profitable rules.

Using a mean-variance optimization framework we develop a simple non linear model of exchange rate with heterogeneous agents and assuming transaction costs in the goods markets. We show that our model is capable of replicating the empirical puzzles and anomalies uncovered in the last decade by the empirical exchange rate literature.

The paper is organized as follows. In sections 1.2 and 1.3 we present the theoretical model. In section 1.4 we present the deterministic solution of the model, while in section 1.5 we analyze its stochastic properties. In section 1.6 we show the empirical relevance of the model. Finally, in section 1.7 we explore the *rationality* of chartists behavior. Section 1.8 concludes.

1.2 The model

In this section we develop a simple non-linear model of the exchange rate. We assume agents of different types i depending on their beliefs about the future exchange rate. Each agent can invest in two assets, a domestic and a foreign one. The agents' utility function can be represented by the following equation:

$$U(W_{t+1}^i) = E_t(W_{t+1}^i) - \frac{1}{2}\mu V^i(W_{t+1}^i) \tag{1.1}$$

where W_{t+1}^i is the wealth of agent of type i at time $t+1$, E_t is the expectation operator, μ is the coefficient of risk aversion and $V^i(W_{t+1}^i)$ represents the conditional variance of wealth of agent i. The wealth is specified as follows:

$$W_{t+1}^i = (1 + r^*)\, s_{t+1} d_t^i + 1 + r\left(W_t^i - s_t d_t^i\right) \tag{1.2}$$

where r and r^* are respectively the domestic and the foreign interest rates, s_{t+1} is the exchange rate at time $t+1$, $d_{i,t}$ represents the holdings of the foreign assets by agent of type i at time t. Thus, the first term on the right-hand side of (1.2) represents the value of the foreign portfolio in domestic currency at time

$t+1$ while the second term represents the value of the domestic portfolio at time $t+1$.

Substituting equation (1.2) in (1.1) and maximizing the utility with respect to $d_{i,t}$ allows us to derive the optimal holding of foreign assets by agents of type i:

$$d_{i,t} = \frac{(1+r^*)\, E_t^i\,(s_{t+1}) - (1+r)\, s_t}{\mu \sigma_{i,t}^2} \tag{1.3}$$

The market demand for foreign assets at time t is the sum of the individual demands, *i.e.*:

$$\sum_{i=1}^{N} n_{i,t} d_{i,t} = D_t \tag{1.4}$$

where $n_{i,t}$ is the number of agents of type i.

Market equilibrium implies that the market demand is equal to the market supply X_t which we assume to be exogenous[3]. Thus,

$$X_t = D_t \tag{1.5}$$

Substituting the optimal holdings into the market demand and then into the market equilibrium equation and solving for the exchange rate s_t yields the equilibrium exchange rate:

$$s_t = \left(\frac{1+r^*}{1+r}\right) \frac{1}{\sum\limits_{i=1}^{N} \frac{n_{i,t}}{\sigma_{i,t}^2}} \left[\sum_{i=1}^{N} n_{i,t} \frac{E_t^i\,(s_{t+1})}{\sigma_{i,t}^2} - \mu \frac{X_t}{1+r}\right] \tag{1.6}$$

In order to model the expectations formation we assume that there are two types of agents: chartists and fundamentalists. As a result equation (1.6) specializes to :

$$s_t = \left(\frac{1+r^*}{1+r}\right) \frac{1}{\left(\frac{n_{f,t}}{\sigma_{f,t}^2} + \frac{n_{c,t}}{\sigma_{c,t}^2}\right)} \left[n_{f,t} \frac{E_t^f\,(s_{t+1})}{\sigma_{f,t}^2} + n_{c,t} \frac{E_t^c\,(s_{t+1})}{\sigma_{c,t}^2} - \mu \frac{X_t}{1+r}\right] \tag{1.7}$$

Thus the exchange rate is determined by the expectations of fundamentalists and chartists about the future exchange rate. These forecasts are weighted by their respective variances.

We now specify how fundamentalists and chartists form their expectations of the future exchange rate. In a second step we will specify how they take into account the risk as measured by the variances.

[3] The market supply is determined by the net current account and by the sales or purchases of foreign exchange of the central bank. We assume both to be exogenous. In an extension of this paper we intend to endogenise the market supply.

The fundamentalists base their forecast on a comparison between the market and the fundamental exchange rate, *i.e.* they forecast the market rate to return to the fundamental rate in the future. In this sense they use a negative feedback rule that introduces a mean reverting dynamics in the exchange rate. The speed with which the market exchange rate returns to the fundamental is assumed to be determined by the speed of adjustment in the goods market. Thus, the forecasting rule for the fundamentalists is:

$$E_t^f\left(\Delta s_{t+1}\right) = -\psi\left(s_t - s_t^*\right) \tag{1.8}$$

where s_t^* is the fundamental exchange rate at time t , which is assumed to follow a random walk.

The chartists forecast the future exchange rate by extrapolating past exchange rate movements. Their forecasting rule can be specified as:

$$E_t^c\left(\Delta s_{t+1}\right) = \beta \sum_{i=0}^{T} \alpha_i \Delta s_{t-i} \tag{1.9}$$

Thus, the chartists compute a moving average of the past exchange rate changes and they extrapolate this into the future exchange rate change. The degree of extrapolation is given by the parameter β. Note that in contrast to the fundamentalists they do not take into account information concerning the fundamental exchange rate. In this sense they can be considered to be pure noise traders. In a way this chartist rule can also be seen as reflecting herding behavior, *i.e.* chartists do not take fundamental information into account because they feel to uncertain about their meaning, but they closely watch the movements of the exchange rate as a way to detect "market sentiments". If the latter are positive, they buy; if they are negative, they sell.

Our choice to give a prominent role to chartists' rules of forecasting is based on empirical evidence. The evidence that Chartism is used widely to make forecasts is overwhelming (see Cheung and Chinn (1989), Taylor and Allen (1992)). It remains important, however, to check if the model is internally consistent. In particular, the chartists' forecasting rule must be shown to be profitable within the confines of the model. If these rules turn out to be unprofitable, they will not continue to be used. We return to this issue when we let the number of chartists be determined by the profitability of the chartists' forecasting rule.

We now analyze how fundamentalists and chartists evaluate the risk. The latter is measured by the variance terms in equation (1.7), which we define as the weighted average of the squared (one period ahead) forecasting errors made by chartists and fundamentalists, respectively. Thus,

$$\sigma_{i,t} = \sum_{k=1}^{\infty} \gamma_k \left[E_{t-k}^i\left(s_{t-k+1}\right) - s_{t-k+1}\right]^2 \tag{1.10}$$

where γ_k are geometrically declining weights.

However fundamentalists and chartists perceive the risk in a different way. In particular the fundamentalists are assumed to take into account the deviation of the exchange rate from the fundamental in addition to the forecasting error. We will call the deviation of the market exchange rate from its fundamental, the misalignment. Thus the fundamentalists' risk term can be written as:

$$\sigma_{f,t} = \frac{\sum_{k=1}^{T} \gamma_k \left[E_{t-k}^{f}\left(s_{t-k+1}\right) - s_{t-k+1}\right]^2}{\left(s_{t-1} - s_{t-1}^{*}\right)^2} \tag{1.11}$$

where $\left(s_{t-1} - s_{t-1}^{*}\right)$ is the misalignment.

The logic behind this specification is that the fundamentalists consider the fundamental exchange rate as a benchmark. The larger is the misalignment the less the fundamentalists will attach importance to the short term volatility as measured by the one-period ahead forecasting error. Put differently, the larger is the deviation of the exchange rate from its fundamental, the smaller is the risk of using a fundamentalist forecasting rule, given the past (short-term) forecast errors made by the fundamentalists. In contrast the chartists do not take into account the misalignment.

We now specify the dynamics that governs the number of chartists and fundamentalists, namely n_{ct} and n_{ft}. In order to do so, we describe how the number of chartists and fundamentalists changes from period $t-1$ to period t:

$$n_t^c = n_{t-1}^c + n_{t-1}^f p_t^{fc} - n_{t-1}^c p_t^{cf} \tag{1.12}$$

$$n_{f,t} = n_{f,t-1} + n_{c,t-1} p_t^{cf} - n_{f,t-1} p_t^{fc} \tag{1.13}$$

where $n_{c,t}$ and $n_{f,t}$ are the number of chartists and fundamentalists in period t; p_t^{cf} represents the probability of a chartist to become a fundamentalist in period t, and p_t^{fc} is the probability of a fundamentalist to become a chartist in period t.

These probabilities are assumed to be a function of the relative utilities of chartism and fundamentalism. These utilities depend on the profitability of the forecasting rules and the risk associated with their use. The probabilities are specified as follows:

$$p_t^{fc} = \varsigma \frac{exp\left(U_{c,t-1}\right)}{exp\left(U_{c,t-1}\right) + exp\left(U_{f,t-1}\right)} \tag{1.14}$$

$$p_t^{cf} = \varsigma \frac{exp\left(U_{f,t-1}\right)}{exp\left(U_{c,t-1}\right) + exp\left(U_{f,t-1}\right)} \tag{1.15}$$

where $U_{c,t-1}$ and $U_{f,t-1}$ are the utility functions of the chartists and fundamentalists in period $t-1$. These probabilities of switching can be interpreted as follows. When the utility associated with using chartist information increases relative to the utility of using fundamentalist information the agents tend to switch from fundamentalism to chartism and vice versa. Taking into account the specification of the utility function (equation (1.1)), we can rewrite equations (1.14) and (1.15) as follows:

$$p_t^{fc} = \varsigma \frac{exp\left(\pi_{c,t-1} - \mu\sigma_{c,t-1}^2\right)}{exp\left(\pi_{c,t-1} - \mu\sigma_{c,t-1}^2\right) + exp\left(\pi_{f,t-1} - \mu\sigma_{f,t-1}^2\right)} \tag{1.16}$$

$$p_t^{cf} = \varsigma \frac{exp\left(\pi_{f,t-1} - \mu\sigma_{f,t-1}^2\right)}{exp\left(\pi_{c,t-1} - \mu\sigma_{c,t-1}^2\right) + exp\left(\pi_{f,t-1} - \mu\sigma_{f,t-1}^2\right)} \tag{1.17}$$

Thus, the probability of switching depends on the relative profits adjusted for the risk. When the risk adjusted profits of chartist rules increase relative to the risk adjusted profits of fundamentalist rules, the probability of switching from fundamentalism to chartism increases, and vice versa. Agents make a profit when they correctly forecast the direction of the exchange rate movements. They make a loss if they wrongly predict the direction of its movements. The profit (the loss) they make equals the one-period return of investing $1. The parameter ς measures the sensitivity of the probability of switching with respect to the relative profits adjusted for the risk. Note also that $0 < \varsigma < 1$.

1.3 The model with transaction costs

There is an increasing body of theoretical literature stressing the importance of transaction costs as a source of non-linearity in the determination of the exchange rate (Dumas (1992), Sercu, Uppal and Van Hulle (1995), Obstfeld and Rogoff (2000)). The importance of transaction costs has also been confirmed empirically (Taylor, Peel, and Sarno (2001), Kilian and Taylor (2001)). Therefore, we will develop a version of the previous model in which the transaction costs play a role.

We take the view that if transaction costs exist, the fundamentalists take this information into account. Therefore, if the exchange rate is within the transaction costs band the fundamentalists behave differently than if the exchange rate moves outside the transaction costs band.

Consider the first case, when the exchange rate deviations from the fundamental value are smaller than the transaction costs. In this case the fundamentalists know that arbitrage in the goods market does not apply. As a result, they expect the changes in the exchange rate to follow a white noise process. The best they can do is to forecast no change. More formally,

$$\text{when} \quad | s_t - s_t^* | < C, \qquad \text{then} \quad E_t^f(\Delta s_{t+1}) = 0.$$

In the second case, when the exchange rate deviation from its fundamental value is larger than the transaction costs C (assumed to be of the 'iceberg' type), then the fundamentalists follow the same forecasting rule as in equation (1.8). More formally,

$$\text{when } | s_t - s_t^* | > C \quad \text{holds,} \quad \text{then equation (1.8) applies.}$$

This formulation implies that when the exchange rate moves outside the transaction costs band, market inefficiencies other than transaction costs continue to play a role. As a result, these inefficiencies prevent the exchange rate from adjusting instantaneously. In our model these inefficiencies are captured by the fact that the speed of adjustment in the goods market is not infinite.

1.4 Solution of the model

In this section we investigate the properties of the solution of the model. We first study its deterministic solution. This will allow us to analyze the characteristics of the solution that are not clouded by exogenous noise. We use simulation techniques since the non-linearities do not allow for a simple analytical solution. We select "reasonable" values of the parameters, *i.e.* those that come close to empirically observed values.

We first concentrate on the fixed point solutions of the model. We find that for a relatively wide range of parameters the solution converges to a fixed point (a fixed-point attractor). However, there are many such fixed points (attractors) to which the solution converges depending on the initial conditions. We illustrate this feature in figure 1.1 where we plot the fixed point solutions (attractors) as a function of the different initial conditions. On the horizontal axis we set out the different initial conditions. These are initial shocks to the deterministic system. The vertical axis shows the solutions corresponding to these different initial conditions. Figure 1.1 shows the fixed point solutions when the speed of adjustment in the goods market is 0.2. Note the complex pattern of these fixed point solutions, with many discontinuities. This has the implication that a small change in the initial condition can have a large effect on the solution. This feature lies at the heart of some of the results that are obtained with this model relating to the unpredictability of the effect of shocks in exogenous variables. We return to this phenomenon in section 1.6.

It should also be noted that the fixed-point attractors lie within the transaction costs band. The intuition is that any fixed-point solution outside the transaction costs band would create an inconsistency, which can be described as follows. Outside the transaction costs band the fundamentalists' behavior leads to a mean reverting process of the exchange rate, moving the latter towards the transaction costs band. Thus, if a fixed point solution was observed outside the transactions cost band, this would mean that the fundamentalists would fail to move the exchange rate towards the band. Once inside the band, the fundamentalists' dynamics disappears. The only dynamics then comes from the chartists who drive the exchange rate to some attractor within the band. The

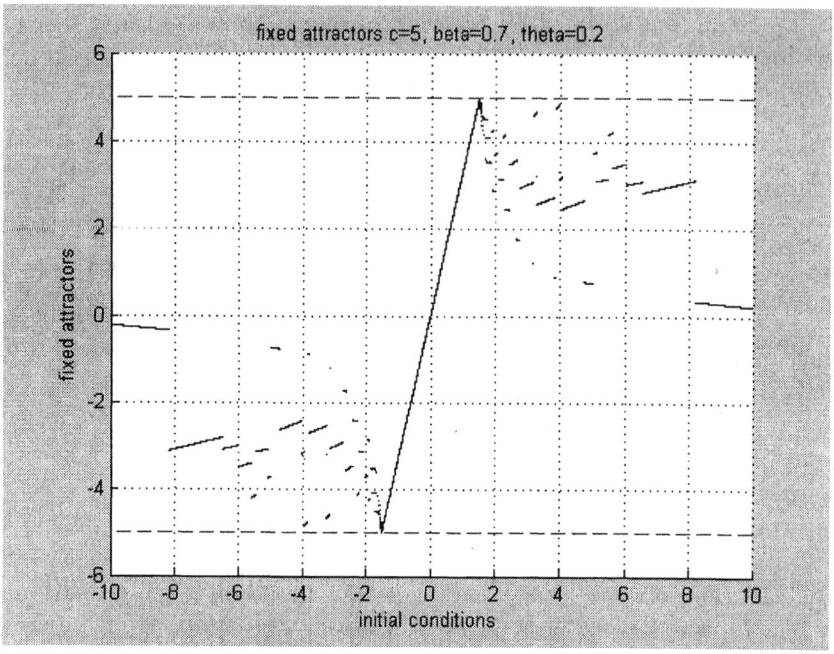

Figure 1.1: Fixed attractors.

exact position of this attractor depends on the entry point of the exchange rate in the transaction costs band, and this depends on the initial shock.

It is also interesting to point out the existence of critical points. These are often obtained close to the limit of the transaction costs band. A continuous line is then broken into a cluster of points. These have a fractal nature, in that they have an infinite self-replicating characteristics.

1.5 The stochastic version of the model

We now introduce stochastic disturbances to the model. In our model these disturbances appear in two places. First they affect the fundamental, which is assumed to be a random walk. Second, as can be seen from equation (1.8), there is exogenous noise leading to forecast errors of chartists and fundamentalists. We simulate the model with a certain combination of parameter values that we refer to as the "standard case". This includes setting $c = 5$, $\beta = 0.5$ and $\theta = 0.2$. (Similar results are obtained for a wide range of parameter values).

A first feature of the solution of the stochastic version of the model is the sensitivity to initial conditions. In order to show this, we first simulated the model with the "standard" parameter values and then we simulated the model with the same parameters setting but with a slightly different initial condition.

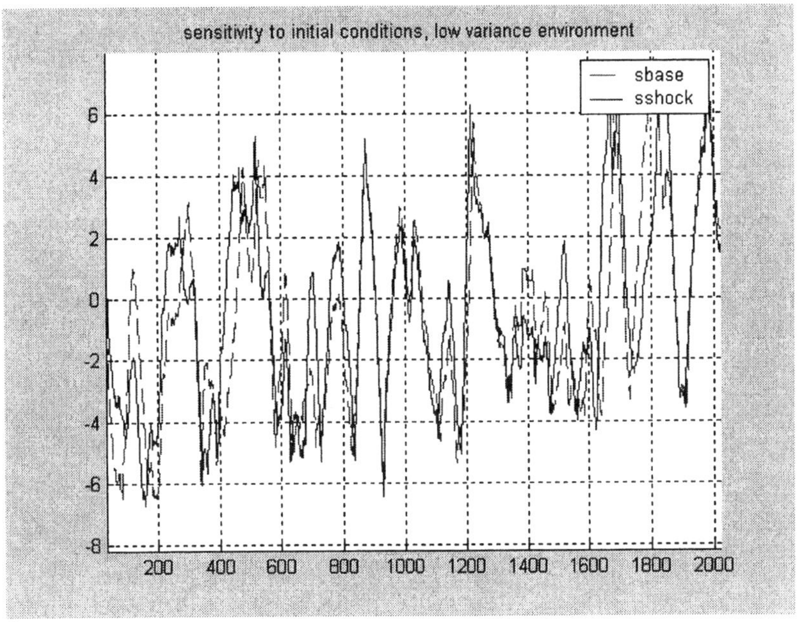

Figure 1.2: Sensitivity to initial conditions.

In both cases we used identical stochastic disturbances. We show the time paths of the (market) exchange rate in figure 1.2.

We observe that after a certain number of periods the two exchange rates start following a different path. This result is related to the presence of many fixed-point attractors in the deterministic part of the model, which are themselves dependent on the initial conditions (see figure 1.1, that shows how slight differences in initial conditions can lead to fixed-point attractors that are very far apart). As a result, the two exchange rates can substantially diverge because they are attracted by fixed-points that are far away from each other. The nice aspect of this is that we obtain a result that is typical for chaotic systems, however, without chaos being present in the deterministic part of the model. The combination of exogenous noise and a multiplicity of fixed-point attractors creates chaos-like dynamics.

A second feature of the model relates to the way shocks in the fundamental exchange rate are transmitted into the market exchange rate. In linear models a permanent shock in the fundamental has a predictable effect on the exchange rate, *i.e.* the coefficient that measures the effect of the shock in the fundamental on the exchange rate converges after some time to a fixed number. Things are very different in our non-linear model. We illustrate this by showing how a permanent increase in the fundamental is transmitted to the exchange rate. We assumed that the fundamental rate increases by 10, and we computed the effect on the exchange rate by taking the difference between the exchange rate with

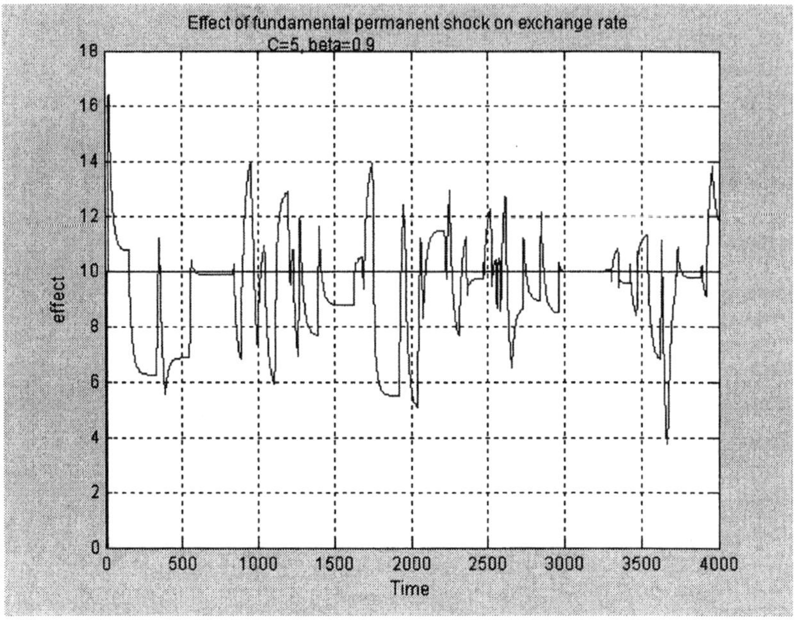

Figure 1.3: Effect of fundamental permanent shock on exchange rate.

the shock and the exchange rate without the shock. In a linear model we would find that in the long run the exchange rate increases by 10. This is not the case in our model. We present the evidence in figure 1.3 where we show the effect of the same permanent shock of 10 in the fundamental rate on the exchange rate. The simulations are done assuming exactly the same stochastics in the scenario with as without the permanent shock in the fundamental exchange rate. Thus, there is no exogenous noise in the model that could blur the transmission process from the fundamental rate to the exchange rate.

The most striking feature of these results is that the effect of the permanent shock does not converge to a fixed number. In fact, it follows a complex pattern. Thus, in a non-linear world it is very difficult to predict what the effect will be of a given shock in the fundamental, even in the long run. Such predictions can only be made in a statistical sense, *i.e.* our model tells us that the effect of a shock of 10 in the fundamental will be to increase the exchange rate by 10 on average. In any given period, however, the effect could deviate substantially from this average prediction.

The importance of the initial conditions for the effect of a permanent shock in the fundamental can also be seen by the following experiment. We simulated the same permanent shock in the fundamental but applied it in two different time periods. In the first simulation we applied the shock in the first period; in the second simulation we applied it in the next period. The exogenous noise

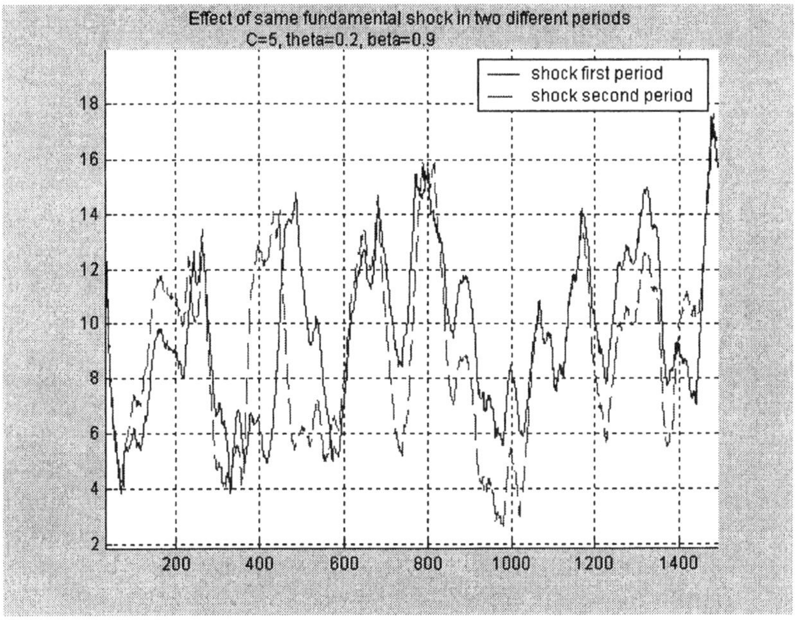

Figure 1.4: Effect of same fundamental shock in two different periods.

was identical in both simulations. Thus the only difference is in the timing of the shock. We show the results in figure 1.4.

We observe that the small difference in timing changes the whole future history of the exchange rate. As a result, the effect of the shock measured at a particular point in time can be very different in both simulations. Thus history matters. The time at which the permanent shock occurs influences the effects of the shock.

Our results help to explain why in the real world it appears so difficult to predict the effects of changes in the fundamental exchange rate on the market rate, and why these effects seem to be very different when applied in different periods. In fact this is probably one of the most intriguing empirical problems. Economists usually explain the difficulty of forecasting the effects of a particular change in one exogenous variable (*e.g.* an expansion of the money stock) by invoking the *ceteris paribus* hypothesis, *i.e.* there are usually other exogenous variables changing unexpectedly, preventing us to isolate the effect of the first exogenous variable. In our model the uncertainty surrounding the effect of a disturbance in an exogenous variable is not due to the failure of the *ceteris paribus* hypothesis. No other exogenous variable is allowed to change. The fact is that the change in the exogenous variable occurs at a particular time, which is different from all other times. Initial conditions (history) matters to forecast the effect of shocks. Since each initial condition is unique, it becomes impossible to forecast the effect of a shock at any given point in time with any precision.

1.6 Empirical relevance of the model

In this section we analyze how well our model mimics some of the empirical anomalies and puzzles that have been uncovered by the flourishing empirical literature. We start with the 'disconnect puzzle'.

1.6.1 The disconnect puzzle

The first and foremost empirical puzzle has been called the "disconnect" puzzle (see Obstfeld and Rogoff (2000)), *i.e.* the exchange rate appears to be disconnected from its underlying fundamentals most of the time. It was first analyzed by John Williamson (1985) who called it the 'misalignment problem'. This puzzle was also implicit in the celebrated Meese and Rogoff studies of the early 1980s documenting that there is no stable relationship between exchange rate movements and the news in the fundamental variables. Goodhart (1989) and Goodhart and Figlioli (1991) found that most of the changes in the exchange rates occur when there is no observable news in the fundamental economic variables. This finding contradicted the theoretical models (based on the efficient market hypothesis), which imply that the exchange rate can only move when there is news in the fundamentals.

Our model is capable of mimicking this empirical regularity. In figure 1.5 we show the market exchange rate and the fundamental rate for a combination of parameters that does not produce deterministic chaos. (Our results hold equally well for a large set of parameter values including those that produce deterministic chaos).

We observe that the market rate can deviate from the fundamental value substantially and in a persistent way. Moreover, it appears that the exchange rate movements are often disconnected from the movements of the underlying fundamental. In fact, they often move in opposite directions.

We show the nature of the disconnect phenomenon in a more precise way by applying a cointegration analysis to the simulated exchange rate and its fundamental using the same parameter values as in figure 1.5 for a sample of 8000 periods. We found that there is a cointegration relationship between the exchange rate and its fundamental. Note that in our setting there is only one fundamental variable. This implies that no bias from omitted variables can occur.

In the next step we specify a EC model in the following way:

$$\Delta s_t = \mu \left(s_{t-1} - \gamma s^*_{t-1} \right) + \sum_{i=1}^{n} \lambda_i \Delta s_{t-i} + \sum_{i=1}^{n} \phi_i \Delta s^*_{t-i} \tag{1.18}$$

The first term on the right hand side is the error correction term. The result of estimating this equation is presented in table 1.1 where we have set $n = 4$.

We find that the error correction coefficient (μ) is very low. This suggests that the mean reversion towards the equilibrium exchange rate takes a very long time. In particular, only 0.5% of the adjustment takes place each period. It

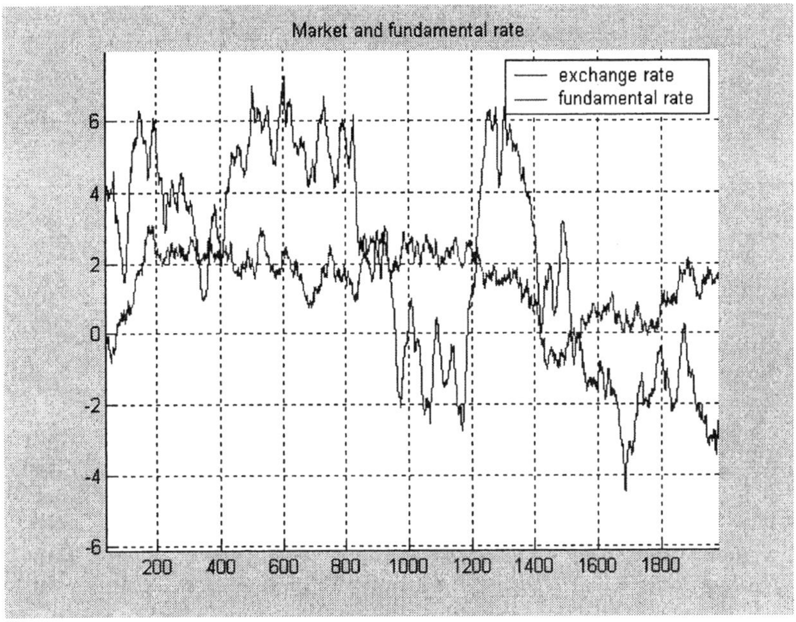

Figure 1.5: Market and fundamental rate.

should be noted that in the simulations we have assumed a speed of adjustment in the goods market equal to 0.2. This implies that each period the adjustment in the goods market is 20%. Thus the nominal exchange rate is considerably slower to adjust towards its equilibrium than what is implied by the speed of adjustment in the goods market. This slow adjustment of the nominal exchange rate is due to the chartists' extrapolation behavior. This phenomenon has been observed in reality. Chen *et al.* (2002) have recently discovered that most of the slow mean reversion of the real exchange rate is due to slow adjustment of the nominal exchange rate and not of the goods prices.

From table 1.1, we also note that the changes in fundamentals have a small and insignificant impact on the change in exchange rate. In contrast, the past changes in the exchange rate play a significant role in explaining the change in exchange rate. These results are consistent with the empirical findings using VAR approach, which suggests that the exchange rate is driven by its own past (see De Boeck (2000))[4].

Thus, our model generates an empirical regularity (the 'disconnect' puzzle) that has also been observed in reality. We can summarize the features of this

[4]We also performed a cointegration analysis for shorter sample periods (1000 periods). We find that in some sample periods the exchange rate and its fundamental are cointegrated, in other sample periods we do not find cointegration. This is in line with the empirical evidence indicating that in some periods the exchange rate seems to be disconnected from its fundamental while in other periods it tightly follows the fundamentals.

Table 1.1: Parameter estimates of EC model (equation (1.18))

EC term		Δs_{t-i}				Δs^*_{t-i}			
μ	γ	λ_1	λ_2	λ_3	λ_4	ϕ_1	ϕ_2	ϕ_3	ϕ_4
−0.005	1.06	0.18	0.15	0.09	0.05	0.04	0.03	0.02	0.01
−6.4	11.3	16.4	13.1	7.06	4.3	2.2	1.6	1.3	1.06

puzzle as follows. First, over the very long run the exchange rate and its fundamentals are cointegrated. However, the speed with which the exchange rate reverts to its equilibrium value is very slow. Second, in the short run the exchange rate and its fundamentals are "disconnected", *i.e.* they do not appear to be cointegrated. Our model closely mimics these empirical regularities.

1.6.2 Fat tails

It is well known that the exchange rate changes do not follow a normal distribution. Instead it has been observed that the distribution of exchange rate changes has more density around the mean than the normal and exhibits fatter tails than the normal (see De Vries (2001)). This phenomenon was first discovered by Mandelbrot (1963), in commodity markets. Since then, fat tails and excess kurtosis have been discovered in many other asset markets including the exchange market. In particular, in the latter the returns have a kurtosis typically exceeding 3 and a measure of fat tails (Hill index) ranging between 2 and 5 (see Koedijk, Stork and de Vries (1992), Huisman *et al.* (2002)). It implies that most of the time the exchange rate movements are relatively small but that occasionally periods of turbulence occur with relatively large exchange rate changes. However, it has also been detected that the kurtosis is reduced under time aggregation. This phenomenon has been observed for most exchange rates (Lux (1998), Calvet and Fisher (2002)). We checked whether this is also the case with the simulated exchange rate changes in our model.

The model was simulated using normally distributed random disturbances (with mean = 0 and standard deviation = 1). We computed the kurtosis and the Hill index of the simulated exchange rate returns. We computed the Hill index for 4 different samples of 2000 observations. In addition, we considered three different cut-off points of the tails (2.5%, 5%, 10%). We show the results of the kurtosis and of the Hill index in table 1.2. We find that for a broad range of parameter values the kurtosis exceeds 3 and the Hill index indicates the presence of fat tails. Finally we check if the kurtosis of our simulated exchange rate returns declines under time aggregation. In order to do so, we chose different time aggregation periods and we computed the kurtosis of the time-aggregated exchange rate returns. We found that the kurtosis declines under time aggregation. In table 1.3 we show the results for some sets of parameter values[5]. This suggests that the non-linear dynamics of the model transforms normally

[5] Another empirical regularity of the distribution of exchange returns is its symmetry. We computed the skewness, and we could not reject that the distribution is symmetric.

Table 1.2: Kurtosis and Hill index

Parameter values			Kurtosis	Median Hill index (4 samples 2000 observations) 2.5% tail	5% tail	10% tail
c	β	θ				
5	0.5	0.2	185.4	2.1	2.8	3.1
5	0.7	0.2	66.7	1.7	1.6	1.9
5	0.7	0.05	5.9	4.6	4.3	4.0
5	0.5	0.1	48.8	2.9	3.4	3.4
5	0.5	0.05	8.5	4.2	4.2	4.0

Table 1.3: Kurtosis under time-aggregation

Parameter values			1 period	5 periods	10 periods	25 periods
c	θ	β	returns	returns	returns	returns
5	0.2	1	7.1	4.5	4.0	2.2
5	0.2	0.8	40.3	13.3	10.3	4.6
5	0.2	0.5	182.9	37.9	24.2	9.9
5	0.1	1	3.8	3.2	3.3	2.4
5	0.3	1	8.9	4.9	3.6	2.2

distributed noise in the exchange rate into exchange rate movements with tails that are significantly fatter than the normal distribution and with more density around the mean. Thus our model mimics an important empirical regularity, *i.e.* that exchange rate movements are characterized by tranquil periods (occurring most of the time) and turbulent periods (occurring infrequently).

1.7 Is chartism evolutionary stable?

An important issue is whether chartism survives in our model. Put differently, we ask the question under which conditions chartism is profitable such that it does not disappear. It should be noted that there is a broad literature that shows that technical analysis is used widely, also by large players (see Taylor and Allen (1992), Wei and Kim (1997)).

We investigate this issue by analyzing how chartism evolves under different conditions. In figure 1.6 we show the share of chartists in the market for increasing values of the extrapolation parameter β. We obtained the chartists weights by simulating the model over 10000 periods and computing the weight of the last period. Our first finding is that chartism does not disappear, *i.e.* in all simulations, for many different parameters configurations, we find that the weight attached to chartists never goes to zero. Second, for a wide range of parameter values we find that the chartists' weight in the market fluctuates around a market share, which exceeds 50%. Thus, in all our simulations of the model we find that chartist rules tend to dominate the fundamentalist rules. Third, as the extrapolation parameter β increases the weight of chartists first in-

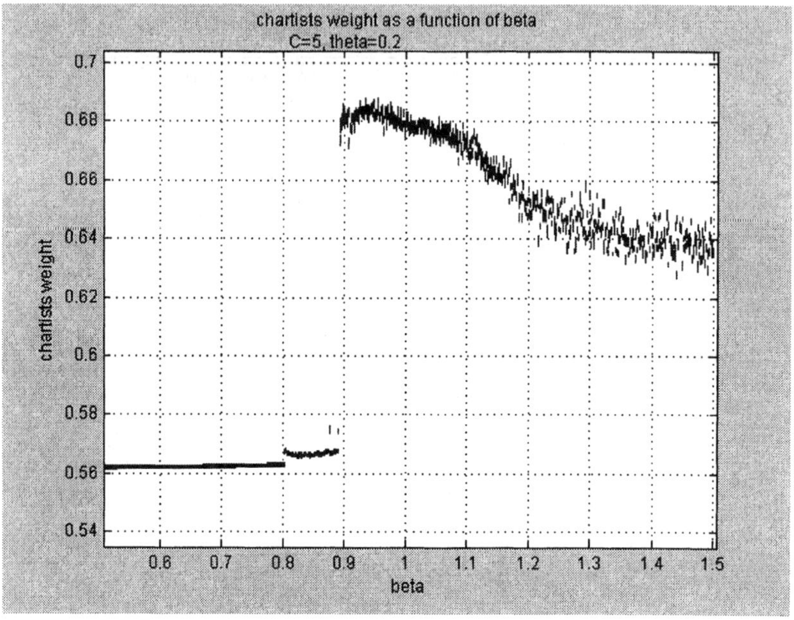

Figure 1.6: Chartists weight as a function of β.

creases slightly. When β reaches a critical point around 0.9, the market share of chartists jumps up to approximately $2/3^6$. It then moves down to settle around 64%. This suggest that as chartists become more aggressive in extrapolating past movements of the exchange rate, the chartist rule becomes more profitable attracting more agents to become chartists. Too much extrapolation in turn reduces the relative profitability of chartism and reduces its market share. The latter result suggests that when chartists extrapolate too much the ensuing deviations from the fundamental exchange rate makes fundamentalist forecasting rules more profitable again.

A final step in the analysis consisted in analyzing the question of how the profits of the chartists and the fundamentalists evolve when the share of the chartists increase exogenously. In order to answer this question we simulated a different version of the model assuming that the shares of chartists vary exogenously from 0 to 0.9 (instead of being determined endogenously by relative profitability). We then computed the average profits of chartists rules as the share of the chartists increases. We did the same for the fundamentalists profits. We show the results in figures 1.7 and 1.8. The results are quite striking. As the weight of chartists increases the profits they make increase strongly. In fact the increase becomes exponential as the share of the chartists in the market becomes very high. In contrast, the profitability of fundamentalists rules is relatively in-

[6] In De Grauwe and Grimaldi (2002) it is shown that this critical point corresponds to the point where the system moves from fix-point solutions into chaos.

sensitive to the number of chartists as long as these are not too numerous. All this suggests that there is a self-fulfilling evolutionary dynamics present in the system which can be described as follows. As the chartists increase in numbers, their profits tend to increase exponentially. The reason is that the increasing importance of chartists increases the noise in the exchange rate movements, making chartism more profitable. At the same time, the chartists have the effect of "creating smoke around the fundamentals", making fundamentalists' forecasting riskier. Another way to interpret this result is that chartism creates noisy information that becomes the source of profitable speculation. The more chartists there are the more such information is created and the more profitable chartists forecasting becomes. Thus, chartists create an informational environment which makes it rational to use chartists' rules.

Why doesn't all this not lead to a corner solution, *i.e.* a situation in which chartism drives out all fundamentalists? As we have seen in the previous paragraphs, the share of the chartists in the market is not driven to 1; it always settles below 1. The reason has to do with risk. When the weight of chartists increases in the market, so does volatility. We show this in figure 1.9 where we present the exchange rate as a function of the weight of chartists. We observe that an increase in the chartists weight leads to a significant increase in volatility especially when the weight approaches 1. Thus, as the weight of chartists in the market increases both profitability and risk of using chartist rules increase. The increasing risk is strong enough to prevents the chartists from completely driving out the fundamentalists and taking over the market.

1.8 Conclusion

In this paper we use a mean variance optimization framework to develop a simple non linear exchange rate model with transaction costs and with heterogeneous agents. Transactions costs produce an important non-linearity in the model. Agents are heterogenous in that they have different beliefs, *i.e.* they use two different forecasting rules. The first type of agents (the chartists) extrapolate past exchange rate movements. They use a positive feedback rule, which tends to destabilize the market. The second type of agents (the fundamentalists) expect the exchange rate to return to its fundamental value. They use a negative feedback rule, which is stabilizing. The relative importance of these two types of agents is assumed to be driven by the relative profitability of their forecasting rules and by the risk associated with the use of such rules. This simple model is capable of creating a remarkably complex dynamics.

The model generates a multitude of fixed-point attractors depending on the initial conditions. Put differently, for each initial condition there is a unique solution. By adding exogenous noise the model produces a complex dynamics. This feature has interesting implications. First, there is sensitivity to initial conditions, which implies that a small disturbance can drive the exchange rate on a different path. Second, the effect of a permanent shock in the fundamental exchange rate has a complex structure. This implies that the effect of a

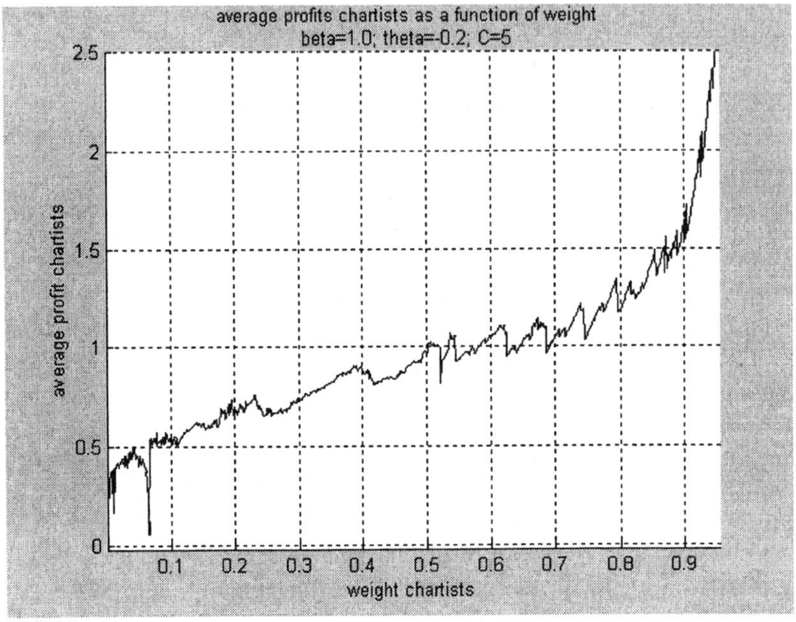

Figure 1.7: Average profits chartists as a funtion of weight.

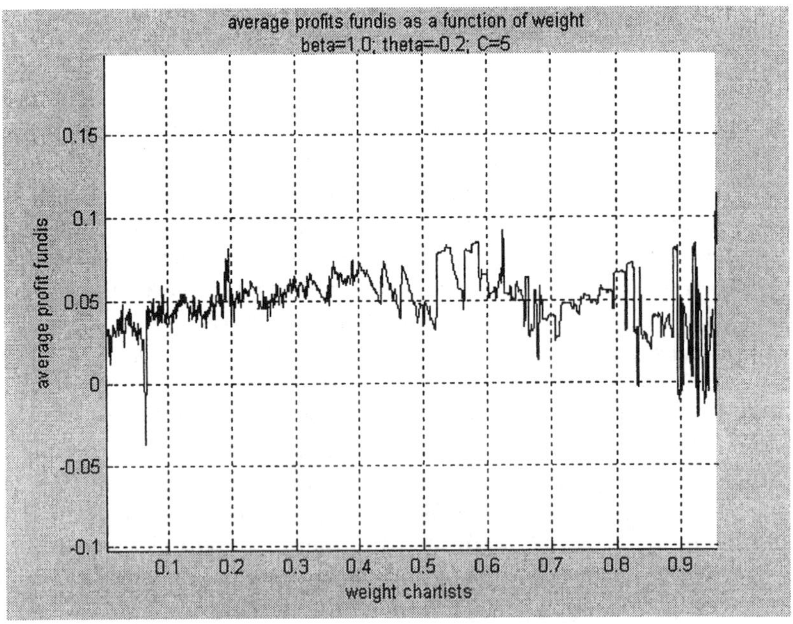

Figure 1.8: Average profits fundamentalists as a funtion of weight.

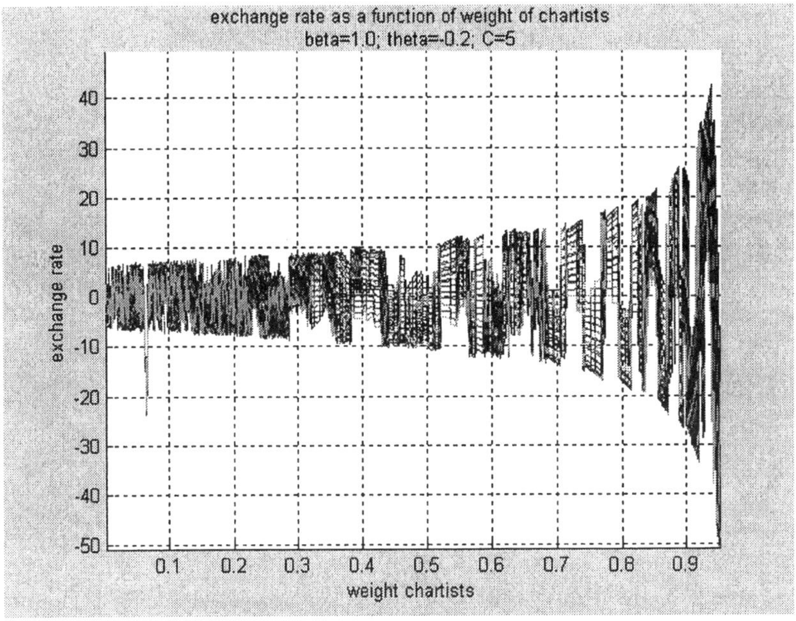

Figure 1.9: Exchange rate as a function of weight of chartists.

permanent shock in the fundamentals is largely unpredictable, *i.e.* one cannot forecast how the shock will affect the exchange rate in any particular point of time, but one can predict the average effect. We also find that the effect of such a shock depends on the exact timing of its occurrence. Thus, history matters. The market has a memory. This contrasts with the prevailing exchange rate models based on the efficient market and rational expectations assumptions that tend to be a-historical.

The empirical relevance of the model is a measure of its quality. Therefore, we analyze to what extent our model is capable of reproducing the exchange rate puzzles that we observe in reality. The first puzzle we analyze is the "disconnect puzzle". This puzzle relates to the fact that the exchange rate movements are disconnected, most of the time, from the movements of the underlying fundamental variables. In our model "disconnection" is a natural outcome of the complex dynamics produced by the interactions between agents using different pieces of information.

Second, fat tails and excess kurtosis, which have been detected in the exchange rate returns, are generated by our model. In other words, our model produces a complex dynamics of the exchange rate with intermittency of periods of high and low turbulence. We find that this alternation of periods of tranquility and turbulence is itself unpredictable.

Finally, we investigated under what conditions chartism is evolutionary stable. We found that chartism does not disappear, *i.e.* in all simulations for many different parameters configurations we find that the number of chartists never goes to zero. This result is consistent with the empirical evidence of the importance of chartism in foreign exchange market. We also detected a self-fulfilling character of chartist profitability, *i.e.* when more chartists enter the market they create more noise and thereby make chartists rules more profitable, inducing more entry. Another way to interpret this result is that chartism creates noisy information that becomes the source of profitable speculation. The more chartists there are the more such information is created and the more profitable chartists forecasting becomes. Thus, chartists create an informational environment which makes it rational to use chartists' rules. This process is stopped, however, because of increasing risk generated by the increased noise that the same chartists produce.

References

Bacchetta, P. and van Wincoop, E. (2003), "Can information heterogeneity explain the exchange rate determination puzzle?", *NBER working paper* 9498.

Baxter, M. and Stockman, A. (1989), "Business cycles and the exchange rate regime. Some international evidence", *Journal of Monetary Economics*, 23, 377-400.

Brock, W. and Hommes, C. (1998), "Heterogeneous beliefs and routes to chaos in a simple asset pricing model", *Journal of Economic Dynamics and Control*, 22, 1235-1274.

Cheung, Y. and Lai, K. (2000), "On the purchasing power parity puzzle", *Journal of International Economics*, 52.

Cheung Y., Lai, K. and Bergman, M. (2001), "Dissecting the PPP puzzle: the unconventional roles of nominal exchange rate and price adjustments", Paper presented at CES-Ifo Conference Munich 2002.

Copeland, L. (2000), *Exchange rates and international finance*, 3rd ed., Prentice Hall.

De Boeck, J. (2000), "The effect of macroeconomic 'news' on exchange rates: a structural VAR approach", *mimeo*, University of Leuven.

De Grauwe, P., Dewachter, H. and Embrechts, M. (1993), *Exchange rate theories. Chaotic models of the foreign exchange markets*, Blackwell.

De Grauwe, P. and Grimaldi, M. (2001), "Exchange rates, prices and money: A long run perspective", *International Journal of Finance and Economics*, 6(4), 289-314.

De Grauwe, P. and Vansteenkiste, I. (2001), "Exchange rates and fundamentals. A non-linear relationship?", *CESifo Working Paper*, no. 577, October.

De Vries, C. (2000), "Fat tails and the history of the guilder", *Tinbergen Magazine*, 4, Fall, 3-6.

De Long, J., Bradford, B., Schleiffer and Summers, L. (1990), "Noise trader risk in financial markets", *Journal of Political Economy.*

Dornbusch, R. (1976), "Expectations and exchange rate dynamics", *Journal of Political Economy,* 84.

Dumas, B. (1992), "Dynamic equilibrium and the real exchange rate in a spatially separated world", *Review of Financial Studies,* 5 (2) , 153-180.

Engel, C. (2000), "Long run PPP may not hold after all", *Journal of International Economics*, 57.

Engel, C. and Morley, J. (2001), "The adjustment of prices and the adjustment of the exchange rate", Discussion paper, Department of Economics, University of Wisconsin.

Evans, M. and Lyons, R. (1999), "Order flow and exchange rate dynamics", *NBER Working Paper*, no. 7317.

Fischer, S. (1982), "Relative prices and inflation in the United States and Germany", *European Economic Review*, 18.

Flood, R. and Rose, A. (1995), "Fixing the exchange rate regime: A virtual quest for fundamentals", *Journal of Monetary Economics*, 36, 3-37.

Frankel, J. and Froot, K. (1986), "The Dollar as a speculative bubble: A tale of fundamentalists and chartists", *NBER Working Paper*, no. 1963.

Frenkel, J. (1981), "The collapse of purchasing power parity during the 1970s", *European Economic Review,* 7.

Goodhart, C. (1989), "News and the foreign exchange market", *LSE Financial Markets Group Discussion paper*, 71.

Goodhart, C. and Figliuoli, L. (1991), "Every minute counts in the foreign exchange markets", *Journal of International Money and Finance*, 10, 23-52.

Guillaume, D. (1996), "Chaos, randomness and order in the foreign exchange markets", PhD Thesis, K.U. Leuven

Hallwood, P. and MacDonald, R. (1994), *International money and finance*, 2nd ed., Blackwell, Oxford.

Huisman, R., Koedijk, K., Kool, C. and Palm, F. (2002), "The tail-fatness of FX returns reconsidered", *in DE Economist*, 150, no. 3, September, 299-312.

Isard, P. (1995), *Exchange rate economics*, Cambridge University Press.

Kilian, L. and Taylor, M. (2001), "Why is it so difficult to beat the random walk forecast of exchange rates?" *Mimeo*, University of Warwick.

Kurz, M. (1994), "On the structure and diversity of rational beliefs", *Economic Theory*, 4, 877-900.

Kurz, M. and Motolese, M. (2000), "Endogenous uncertainty and market volatility", *mimeo*, Stanford University.

Li, K. (1999), "Testing symmetry and proportionality in PPP: A panel data approach", *Journal of Business and Economic Statistics,* 17 (4) , 409-418.

Lux, T. (1998), "The socio-economic dynamics of speculative markets: interacting agents, chaos, and fat tails of return distributions", *Journal of Economic Behaviour and Organisation*, 33.

Lux, T. and Marchesi, M. (2000), "Volatility clustering in financial markets: a microsimulation of interacting agents", *International Journal of Theoretical and Applied Finance.*

Lyons, R. (2001), *The microstructure approach to exchange rates*, MIT Press, Cambridge, Mass.

Mandelbrot, B. (1963), "The variation of certain speculative prices", *The Journal of Business*, 36, 394-419.

Meese, R. and Rogoff, K. (1983), "Empirical exchange rate models of the Seventies: Do they fit out of sample?", *Journal of International Economics*, 14, 3-24.

Michael, P., Nobay, R. and Peel, A. (1997), "Transaction costs and non-linear adjustment in real exchange rates: an empirical investigation", *Journal of Political Economy,* 105 (4), 862-879.

Obstfeld, M. and Rogoff, K. (1996), *Foundations of international macroeconomics*, MIT Press, Cambridge, Mass.

Obstfeld, M., and Rogoff, K. (2000), "The six major puzzles in international macroeconomics: Is there a common cause?", *NBER Working Paper* no. 7777, July.

Okun, A. (1971), "The mirage of steady inflation", *Brooking Papers on Economic Activity.*

Rogoff, K. (1996), "The purchasing power parity puzzle", *Journal of Economic Literature*, 34, June, 647-668.

Schittenkopf, C., Dorffner, G. and Dockner, E. (2001), "On nonlinear, stochastic dynamics in economics and financial time series", *Studies in Nonlinear Dynamics and Econometrics,* 4(3), 101-121.

Schleiffer, A. (2000), *Introduction to behavioural finance*, Clarendon Press.

Taylor, M. and Allen, H. (1992), "The use of technical analysis in the foreign exchange market", *Journal of International Money and Finance*, 11, 304-14.

Taylor, M., Peel, D. and Sarno, L. (2001), "Non-linear mean reversion in real exchange rates: towards a solution to the purchasing power parity puzzles", *CEPR discussion paper* no 2658.

Wei, S-J. and Kim, J. (1997), "The big players in the foreign exchange market:do they trade on information or noise?", *NBER working paper* 6256.

Williamson, J. (1985), "The exchange rate system", *Policy Analyses in International Economics*, 5, Institute for International Economics, Washington, D.C.

Chapter 2

Real exchange rate persistence in DGE sticky-price models: An analytical characterization

By Hafedh BOUAKEZ[1]

Abstract

This paper assesses analytically the ability of dynamic general-equilibrium sticky-price models to generate persistent real exchange rate fluctuations. It develops a tractable general-equilibrium model with Calvo-type price stickiness. The model has a closed-form solution and the persistence of the real exchange rate is explicitly characterized. The paper shows that real exchange rate persistence is pinned down by the probability of not changing prices. This result suggests that standard sticky-price models are unable to generate endogenous persistence.

2.1 Introduction

Examining real exchange rate movements within dynamic general-equilibrium (DGE) sticky-price models has been a major development in the international

[1]International Department, Bank of Canada, 234 Wellington St., Ottawa, Canada K1A OG9. Tel.: 1-613-782-8364; Fax: 1-613-782-7658. Email: hbouakez@bank-banque-canada.ca. The views expressed in this paper are mine. No responsibility for them should be attributed to the Bank of Canada.
I would like to thank Steve Ambler, Jeannine Bailliu, Emanuela Cardia, Roger Farmer, Césaire Meh, Kevin Moran, Larry Schembri, Takashi Kano, and Nooman Rebei for helpful comments and discussions.

finance literature over the past decade. This approach has become popular because it reformulates the intuitive arguments of Keynesian theory within a neoclassical micro-founded intertemporal framework. The first generation of DGE sticky-price models of exchange rate determination includes those by Obstfeld and Rogoff (1995) and Betts and Devereux (1996). In both studies, a closed-form solution of the model is made possible by assuming that prices are held fixed for only one period. Because such short-lived price rigidity is obviously insufficient to generate persistent effects of monetary shocks on the real exchange rate, subsequent studies that build on Obstfeld and Rogoff's work endeavored to allow for richer mechanisms of price setting. For example, Bergin and Feenstra (2001) and Chari, Kehoe, and McGrattan (2002) assume staggered price contracts à la Taylor, while Kollmann (2001) introduces a Calvo-type price setting. These pricing mechanisms are certainly more plausible than the one-period price stickiness, but this gain in realism came at the cost of precluding analytical solutions, so that numerical methods were needed to solve the model[2].

Does price staggering generate persistent real exchange rates, once incorporated in a DGE open-economy framework? This question has been recently investigated by Chari, Kehoe, and McGrattan (2002). Using a simulation-based approach, they show that, unless one assumes an implausibly long duration of price contracts, standard DGE sticky-price models fail to match the persistence found in real exchange rate data.

This paper adopts an analytical approach rather than a simulation-based approach to assess the ability of sticky-price models to generate persistent real exchange rates. The paper develops a tractable DGE two-country sticky-price model that can be solved explicitly. The model embeds the standard features found in a typical DGE sticky-price model, yet is simple enough to yield a closed-form solution. The type of price stickiness introduced follows Calvo (1983) where each firm has a constant probability of changing its price in every period.

The paper shows analytically that real exchange rate persistence is pinned down by the probability of not changing prices. More precisely, the real exchange rate response to a monetary shock dies out *exactly* at the same rate at which prices adjust. An interpretation of this result is that the effects of monetary shocks on the real exchange rate do not last beyond the average duration of price contracts. Thus, price staggering does not generate endogenous real exchange rate persistence. Using Taylor's terminology, this means that the *contract multiplier* is exactly equal to 1. This finding parallels the *persistence problem* emphasized by Chari, Kehoe, and McGrattan (2000) who show that, in a closed-economy set-up, staggered price contracts do not generate endogenous output persistence.

The rest of the paper is organized as follows. Section 2.2 presents the theoretical model. Section 2.3 analyzes the model's dynamics. Section 2.4 performs a robustness analysis. Section 2.5 concludes.

---

[2] Bergin and Feenstra (2001) do find an analytical solution to their model in the special case where price contracts last exactly two periods. For longer durations, however, their model cannot be solved analytically.

2.2 The model

The model consists of two countries, each characterized by (i) a representative infinitely lived household, (ii) a representative final-good producer, (iii) a continuum of intermediate-good producers indexed by $i \in [0,1]$, and (iv) a government. A fraction n (respectively, $1-n$) of intermediate-good producers are located in the home (foreign) country. Intermediate goods are differentiated and are used to produce the final good in both countries. The final good is used exclusively for consumption and is not tradable between the two countries. Following Chari, Kehoe, and McGrattan (2002), I assume that, in each period t, the world economy experiences one of finitely many events s_t. The history of events realized up to period t (or the state of nature in period t) is denoted by $s^t = (s_0, s_1, ..., s_t)$. The probability, as of period 0, of any particular state s^t is denoted by $\omega(s^t)$. The initial state, s^0, is given.

2.2.1 Households

The representative household in the home country has the following lifetime utility function:

$$U_0 = \sum_{t=0}^{\infty} \sum_{s^t} \beta^t \omega(s^t) u(c(s^t), m(s^t), l(s^t)),$$

where β is the subjective discount factor $(0 < \beta < 1)$, and u is the instantaneous utility function. Households derive utility from consumption (c), from holding real money balances (m), and from leisure $(1-l)$[3]. The instantaneous utility function is assumed to be

$$u(c(s^t), m(s^t), l(s^t)) = v(c(s^t), l(s^t)) + \gamma \log(m(s^t)), \tag{2.1}$$

where the function v satisfies $v_c > 0, v_l < 0, v_{cc} < 0$, and $v_{ll} > 0$, $m = M/P$, M is the nominal money stock, P is the aggregate price index, and γ is a positive parameter.

Financial markets are assumed to be complete. That is, there exists a complete set of state-contingent bonds, which are assumed to be denominated in domestic currency. Let $B(s^{t+1})$ denote the home household's holdings of a bond purchased in period t that pays one unit of the home currency in period $t+1$ if state s^{t+1} occurs and 0 otherwise. The price of this bond in units of the home currency is denoted by $Q(s^{t+1}|s^t)$.

The representative household carries $M(s^{t-1})$ units of home currency and a portfolio of state-contingent nominal bonds into period t. After the realization of the event s_t, the household receives $B(s^t)$ additional units of home currency. It also receives a lump-sum transfer, $T(s^t)$, from the government and dividends, $D(i, s^t)$, from each intermediate-good producer $i \in [0, n]$. The household sells

[3] In each period, the household's total endowment of time is normalized to unity.

$l(i, s^t)$ units of labor to each intermediate-good producer $i \in [0, n]$ at the nominal wage, $W(s^t)$. The household's income in period t is allocated to consumption, money holdings and the purchase of nominal bonds. The representative household's budget constraint in period t is

$$P(s^t)c(s^t) + \sum_{s^{t+1}} Q(s^{t+1}|s^t)B(s^{t+1}) + M(s^t)$$
$$\leq B(s^t) + M(s^{t-1}) + W(s^t)l(s^t) + D(s^t) + T(s^t), \qquad (2.2)$$

where $l(s^t) = \int_0^n l(i, s^t)di$ is the household's total labor supply, and $D(s^t) = \int_0^n D(i, s^t)di$ are total dividends.

The representative household in the foreign country has the following budget constraint:

$$P^*(s^t)c^*(s^t) + \sum_{s^{t+1}} Q(s^{t+1}|s^t)B^*(s^{t+1})/e(s^t) + M^*(s^t)$$
$$\leq B^*(s^t)/e(s^t) + M^*(s^{t-1}) + W^*(s^t)l^*(s^t) + D^*(s^t) + T^*(s^t), \qquad (2.3)$$

where the asterisk denotes variables in the foreign country, and e_t is the nominal exchange rate, defined as the price of one unit of the foreign currency in terms of the home currency.

The representative household maximizes its lifetime utility subject to its budget constraint (2.2). The first-order necessary conditions for this problem are

$$v_l(s^t) + w(s^t)v_c(s^t) = 0, \qquad (2.4)$$
$$\gamma/m(s^t) - v_c(s^t) + \beta \sum_{s^{t+1}} \omega(s^{t+1}|s^t)v_c(s^{t+1})/\pi(s^{t+1}) = 0, \qquad (2.5)$$
$$Q(s^{t+1}|s^t)v_c(s^t) - \beta\omega(s^{t+1}|s^t)v_c(s^{t+1})/\pi(s^{t+1}) = 0, \qquad (2.6)$$

where $w(s^t)$ is the real wage, $\omega(s^{t+1}|s^t) = \omega(s^{t+1})/\omega(s^t)$ is the conditional probability of s^{t+1} given s^t, and $\pi(s^{t+1}) = P(s^{t+1})/P(s^t)$ is the gross inflation rate between t and $t+1$.

The foreign household's problem implies an analogous set of first-order conditions. Equation (2.6) and its foreign counterpart imply the following risk-sharing condition:

$$q(s^t) = \varsigma v_c^*(s^t)/v_c(s^t), \qquad (2.7)$$

where $q = eP^*/P$ is the real exchange rate, and ς is a constant reflecting initial wealth differences.

From this point on, I will use the operator E_t to denote the expected value of variables dated $\tau \geq t$ conditional on the current state s^t. More specifically, for a given variable x, $E_t x(s^\tau) = \sum_{s^\tau} \omega(s^\tau|s^t)x(s^\tau)$. Moreover, for ease of presentation, I will slightly abuse the notation by writing $x_t \equiv x(s^t)$[4].

[4]The first-order condition (2.5), therefore, becomes $\gamma/m_t - v_{ct} + \beta E_t(v_{ct+1}/\pi_{t+1}) = 0$.

2.2.2 The final-good producer

Final-good producers are perfectly competitive. They use the differentiated intermediate goods from both countries to produce a single country-specific perishable commodity using the following CES technology:

$$y_t = \left[\int_0^n y_{ht}(i)^{(\theta-1)/\theta} di + \int_n^1 y_{ft}(i)^{(\theta-1)/\theta} di \right]^{\theta/(\theta-1)}, \tag{2.8}$$

where $y_{ht}(i)$ (respectively, $y_{ft}(i)$) is the input of intermediate good i produced in the home (foreign) country, and $\theta > 1$ is the elasticity of substitution between different goods. As $\theta \to \infty$, goods become perfect substitutes in production. It is assumed that exports are invoiced in the currency of the importing country. This assumption, often called local currency pricing, was introduced by Betts and Devereux (1996, 2000) into Obstfeld and Rogoff's (1995) model to characterize pricing-to-market behavior by monopolistic firms. Pricing-to-market is the ability of a monopoly to set different prices in the home and foreign countries by somehow segmenting the market. Typically, this price discrimination leads to the violation of the law of one price among traded goods, and ultimately to a departure from the purchasing power parity. It is clear, though, that such behavior is possible only if there are economic and/or institutional constraints that prevent consumers from taking advantage of international arbitrage opportunities in the goods market. Empirically, studies by Knetter (1989, 1993), Engel (1993), and Engel and Rogers (1996) seem to provide strong evidence in favor of pricing-to-market, as departures from purchasing power parity were found to reflect mainly the failure of the law of one price between traded goods, rather than the presence of non-traded goods. Under the assumption of local currency pricing, the final-good producer solves the following problem:

$$\underset{\{y_{ht}(i), y_{ft}(i)\}}{Max} \qquad P_t y_t - \textstyle\int_0^n P_{ht}(i) y_{ht}(i) di + \int_n^1 P_{ft}(i) y_{ft}(i) di,$$

subject to (2.8), where $P_{ht}(i)$ (respectively, $P_{ft}(i)$) is the price of intermediate good i produced in the home (foreign) country. The solution of this problem yields the input demand of good i:

$$y_{jt}(i) = (P_{jt}(i)/P_t)^{-\theta} y_t, \tag{2.9}$$

where $j = h$ for $i \in [0, n]$ and $j = f$ for $i \in]n, 1]$, and the elasticity of demand is equal to θ. The zero-profit condition implies that the aggregate price index is given by

$$P_t = \left[\int_0^n P_{ht}(i)^{1-\theta} di + \int_n^1 P_{ft}(i)^{1-\theta} di \right]^{1/(1-\theta)}. \tag{2.10}$$

Let P_{ht} and P_{ft} denote, respectively, the price indexes of home and foreign intermediate goods sold in the home country[5]. Hence, the aggregate price index can be written as

$$P_t = \left[n P_{ht}{}^{1-\theta} + (1-n) P_{ft}{}^{1-\theta} \right]^{1/(1-\theta)}. \quad (2.11)$$

The problem of the representative foreign final good producer is described in an analogous manner.

2.2.3 The intermediate-good producer

The representative firm i in the home country produces its differentiated good using the following technology:

$$y_t(i) \equiv y_{ht}(i) + y_{ht}^*(i) = F(k_t(i), h_t(i)),$$

where $k_t(i)$ and $h_t(i)$ are, respectively, capital and labor input used in the production of the intermediate good i, and F is a concave production function with constant returns to scale[6]. Denoting by r_t the real rental rate of capital, the representative final-good producer solves the following problem:

$$\underset{\{k_t(i), h_t(i)\}}{Min} \quad r_t k_t(i) + w_t h_t(i),$$

subject to $F(k_t(i), h_t(i)) = 1$. First-order conditions are

$$r_t = mc_t F_{kt}, \quad (2.12)$$

$$w_t = mc_t F_{ht}, \quad (2.13)$$

where the real marginal cost, mc_t, is the Lagrange multiplier associated with the constraint. Using equations (2.12) and (2.13), and the fact that the function F is homogenous of degree 1, it is straightforward to show that the total real cost for firm i is equal to $mc_t(y_{ht}(i) + y_{ht}^*(i))$.

Intermediate-good producers are monopolistically competitive. Each firm faces a downward-sloping demand curve for its differentiated good in each country. Firm i chooses its (nominal) prices, $P_h(i)$ and $P_h^*(i)$, taking as given the aggregate demand and the price level in each country. Nominal prices are assumed to be sticky. Price stickiness is modeled *à la* Calvo (1983). That is, each period, some firms are randomly selected to set new prices for the home and foreign markets. The probability of being selected in any particular period is constant and is equal to $1-\varphi$.

Let $\tilde{P}_{ht}$ and $\tilde{P}_{ht}^*$ denote the optimal prices set by a typical firm at period t in the home and foreign countries, respectively. It is not necessary to index $\tilde{P}_{ht}$ and $\tilde{P}_{ht}^*$ by firm, because all of the firms that change their prices at a given

---

[5] More precisely, P_{ht} and P_{ft} are defined as follows: $P_{ht} \equiv \left[\frac{1}{n}\int_0^n P_{ht}^{1-\theta}(i)di\right]^{1/(1-\theta)}$ and $P_{ft} \equiv \left[\frac{1}{1-n}\int_n^1 P_{ft}^{1-\theta}(i)di\right]^{1/(1-\theta)}$.

[6] Labor-market clearing requires that $\int_0^n h_t(i)di = nl_t$.

time choose the same price. The total domestic and foreign demands facing this firm at time τ for $\tau \geq t$ are $\tilde{y}_{h\tau} = (\tilde{P}_{ht}/P_\tau)^{-\theta} y_\tau$ and $\tilde{y}^*_{h\tau} = (\tilde{P}^*_{ht}/P_\tau)^{-\theta} y^*_\tau$, respectively. The probability that $\tilde{P}_{ht}$ and $\tilde{P}^*_{ht}$ "survive" at least until period τ, for $\tau \geq t$, is $\varphi^{\tau-t}$. Thus, the intermediate-good producer chooses $\tilde{P}_{ht}$ and $\tilde{P}^*_{ht}$ to maximize

$$E_t \sum_{\tau=t}^{\infty} (\varphi\beta)^{\tau-t} \Lambda_{t,\tau} \left[\tilde{P}_{ht}\tilde{y}_{h\tau} + e_\tau \tilde{P}^*_{ht}\tilde{y}^*_{h\tau} - MC_\tau(\tilde{y}_{h\tau} + \tilde{y}^*_{h\tau}) \right],$$

where $\Lambda_{t,\tau}$ is the marginal utility of a dollar earned at time τ relative to its marginal utility at time t, and $MC_t = mc_t P_t$ is the nominal marginal cost. First-order conditions for this problem are

$$\tilde{P}_{ht} = \frac{\theta}{\theta-1} \frac{E_t \sum_{\tau=t}^{\infty} (\varphi\beta)^{\tau-t} \Lambda_{t,\tau} MC_\tau \tilde{y}_{h\tau}}{E_t \sum_{\tau=t}^{\infty} (\varphi\beta)^{\tau-t} \Lambda_{t,\tau} \tilde{y}_{h\tau}}, \tag{2.14}$$

$$\tilde{P}^*_{ht} = \frac{\theta}{\theta-1} \frac{E_t \sum_{\tau=t}^{\infty} (\varphi\beta)^{\tau-t} \Lambda_{t,\tau} MC_\tau \tilde{y}^*_{h\tau}}{E_t \sum_{\tau=t}^{\infty} (\varphi\beta)^{\tau-t} \Lambda_{t,\tau} e_\tau \tilde{y}^*_{h\tau}}. \tag{2.15}$$

Assuming that price changes are independent across firms, the law of large numbers implies that $1-\varphi$ is also the proportion of firms that set a new price each period. The proportion of firms that set a new price at time τ and have not changed it as of time t (for $\tau \leq t$), is given by the probability that a time-τ price is still in effect in period t. It is easy to show that this probability is $\varphi^{t-\tau}(1-\varphi)$. It follows that P_{ht} and P^*_{ht} can be written respectively as

$$P_{ht} = \left((1-\varphi) \sum_{\tau=-\infty}^{t} \varphi^{t-\tau} \tilde{P}_{h\tau}^{1-\theta} \right)^{\frac{1}{1-\theta}}, \tag{2.16}$$

$$P^*_{ht} = \left((1-\varphi) \sum_{\tau=-\infty}^{t} \varphi^{t-\tau} \tilde{P}_{h\tau}^{*1-\theta} \right)^{\frac{1}{1-\theta}}. \tag{2.17}$$

2.2.4 The government

The government represents both the fiscal and monetary authorities in each country. There is no government spending or investment. Each period, the government makes lump-sum transfers to households. Transfers are financed by printing additional money in each period. Thus, the government budget constraint in the home country is

$$T_t = M_t - M_{t-1}. \tag{2.18}$$

Money is supplied exogenously by the government according to $M_t = \mu_t M_{t-1}$, where μ_t is the gross rate of money growth. In real terms, this process implies

$$m_t \pi_t = \mu_t m_{t-1}. \tag{2.19}$$

The rate of money growth, μ_t, is assumed to be a non-autocorrelated shock that follows the process

$$\ln \mu_t = \ln \mu + \epsilon_t, \tag{2.20}$$

where μ is the steady-state rate of money growth, and ϵ_t is a normally distributed zero-mean disturbance.

2.3 Analytical results

To solve the model, I follow the usual strategy of considering an approximate solution in the neighborhood of the steady state. I do so by log-linearizing the equilibrium conditions around a zero-shock initial steady state in which all variables are constant. The steady state corresponds to a symmetric flexible-price equilibrium. From the log-linearized version of the model, it is easy to show that the real exchange rate (expressed as a percentage deviation from its steady-state value) is fully determined by the following three-equation system (see Appendix A for the derivation):

$$\hat{m}_t - \hat{m}_t^* = \hat{m}_{t-1} - \hat{m}_{t-1}^* - (\hat{\pi}_t - \hat{\pi}_t^*) + \hat{\mu}_t - \hat{\mu}_t^*, \tag{2.21}$$

$$E_t\left(\hat{\pi}_{t+1} - \hat{\pi}_{t+1}^*\right) = \frac{1}{\beta}(\hat{\pi}_t - \hat{\pi}_t^*) - \frac{(1-\varphi)(1-\varphi\beta)}{\varphi\beta}\hat{q}_t, \tag{2.22}$$

$$E_t\hat{q}_{t+1} = \frac{1}{\beta}\hat{q}_t - E_t\left(\hat{\pi}_{t+1} - \hat{\pi}_{t+1}^*\right) - \frac{1-\beta}{\beta}(\hat{m}_t - \hat{m}_t^*), \tag{2.23}$$

where the circumflex denotes the percentage deviation of a variable from its steady-state value $[\hat{x}_t = (x_t - x)/x]$. The log-linearized model (2.21)–(2.23) can be written as

$$E_t\mathbf{x}_{t+1} = \mathbf{A}\mathbf{x}_t + \mathbf{B}z_t, \tag{2.24}$$

where $\mathbf{x}_t = (\hat{m}_{t-1} - \hat{m}_{t-1}^*, \hat{\pi}_t - \hat{\pi}_t^*, \hat{q}_t)'$, $z_t = \hat{\mu}_t - \hat{\mu}_t^*$, and $\mathbf{A}$ and $\mathbf{B}$ are respectively, a 3×3 matrix and a 3×1 vector given by

$$\mathbf{A} = \begin{bmatrix} 1 & -1 & 0 \\ 0 & \frac{1}{\beta} & -\frac{(1-\varphi)(1-\varphi\beta)}{\varphi\beta} \\ -\frac{1-\beta}{\beta} & -1 & \frac{1-\varphi\beta(1-\varphi)}{\varphi\beta} \end{bmatrix},$$

and

$$\mathbf{B} = \begin{bmatrix} 1 \\ 0 \\ -\frac{1-\beta}{\beta} \end{bmatrix}$$

Blanchard and Kahn (1980) derive the necessary and sufficient condition for the existence of a unique stable solution to linear dynamic rational-expectation models similar to (2.24). This condition requires that the number of eigenvalues of the matrix **A** outside the unit circle be equal to the number of non-predetermined variables. This saddle-point property guarantees the existence of unique initial values for the non-predetermined variables such that the model always converges to a stationary equilibrium following a shock[7]. Of the three variables included in the vector $\mathbf{x}_t$, one is predetermined $(\hat{m}_{t-1} - \hat{m}^*_{t-1})$ and two are forward-looking ($\hat{\pi}_t - \hat{\pi}^*_t$ and $\hat{q}_t$). Hence, for the model (2.24) to possess a unique stable solution, it must have one stable and two unstable roots.

Proposition 2.1 *The eigenvalues of the matrix* **A** *are* φ, $1/\beta$, *and* $1/\varphi\beta$.

Proof.

$$det(\mathbf{A} - \delta\mathbf{I}) = \begin{vmatrix} 1-\delta & -1 & 0 \\ 0 & \frac{1}{\beta} - \delta & -\frac{(1-\varphi)(1-\varphi\beta)}{\varphi\beta} \\ -\frac{1-\beta}{\beta} & -1 & \frac{1-\varphi\beta(1-\varphi)}{\varphi\beta} - \delta \end{vmatrix}$$

$$= (1-\delta)\begin{vmatrix} \frac{1}{\beta} - \delta & -\frac{(1-\varphi)(1-\varphi\beta)}{\varphi\beta} \\ -1 & \frac{1-\varphi\beta(1-\varphi)}{\varphi\beta} - \delta \end{vmatrix} + \begin{vmatrix} 0 & -\frac{(1-\varphi)(1-\varphi\beta)}{\varphi\beta} \\ -\frac{1-\beta}{\beta} & \frac{1-\varphi\beta(1-\varphi)}{\varphi\beta} - \delta \end{vmatrix}$$

$$= (1-\delta)\left[\left(\frac{1}{\beta} - \delta\right)\left(\frac{1-\varphi\beta(1-\varphi)}{\varphi\beta} - \delta\right) - \left(\frac{(1-\varphi)(1-\varphi\beta)}{\varphi\beta}\right)\right] - \left(\frac{1-\beta}{\beta}\right)\left(\frac{(1-\varphi)(1-\varphi\beta)}{\varphi\beta}\right)$$

$$= \left(\delta^2 - \frac{1+\varphi^2\beta}{\varphi\beta}\delta + \frac{1}{\beta}\right)\left(\frac{1}{\beta} - \delta\right)$$

$$= (\delta - \varphi)\left(\delta - \frac{1}{\varphi\beta}\right)\left(\frac{1}{\beta} - \delta\right).$$

The roots of the characteristic equation $det(\mathbf{A} - \delta\mathbf{I}) = \mathbf{0}$ are $\delta = \varphi$, $\delta = 1/\beta$, or $\delta = 1/\varphi\beta$, which are therefore the eigenvalues of **A**. ■

The matrix **A** has two eigenvalues outside the unit circle ($1/\beta$, and $1/\varphi\beta$) and one eigenvalue of modulus less than 1 (φ). Thus, a unique and stable solution to (2.24) always exists for any sensible values of the parameters φ and β. The magnitude of the stable eigenvalue (φ) characterizes the speed of convergence of the model's variables to their steady-state values, following a structural perturbation. In particular, the following result holds:

Proposition 2.2 *The deviation of the real exchange rate from its steady-state value evolves according to the following process:*

$$\hat{q}_t = \varphi\hat{q}_{t-1} + \varphi z_t.$$

[7] The Blanchard-Kahn condition rules out the possibility of explosive solutions (which arise in the case where the number of eigenvalues greater than 1 exceed the number of non-predetermined variables) and sunspot equilibria (which arise in the case where the number of eigenvalues greater than 1 falls short of the number of non-predetermined variables).

Proof.
Equation (2.21) implies that

$$E_t\left(\hat{\pi}_{t+1}-\hat{\pi}^*_{t+1}\right)=\hat{m}_t-\hat{m}^*_t-E_t(\hat{m}_{t+1}-\hat{m}^*_{t+1}). \tag{2.25}$$

Using this equation to substitute for $E_t\left(\hat{\pi}_{t+1}-\hat{\pi}^*_{t+1}\right)$ in equation (2.23), gives

$$E_t\hat{q}_{t+1}=\frac{1}{\beta}\hat{q}_t+E_t(\hat{m}_{t+1}-\hat{m}^*_{t+1})-\frac{1}{\beta}\left(\hat{m}_t-\hat{m}^*_t\right).$$

The solution to this first-order stochastic difference equation is

$$\hat{q}_t=\hat{m}_t-\hat{m}^*_t. \tag{2.26}$$

Substituting for $\hat{\pi}_t-\hat{\pi}^*_t$ and $E_t\left(\hat{\pi}_{t+1}-\hat{\pi}^*_{t+1}\right)$ in (2.22) using (2.21), (2.25), and (2.26), yields the following second-order stochastic difference equation:

$$E_t\hat{q}_{t+1}-\left(\frac{1+\beta}{\beta}+\frac{(1-\varphi)(1-\varphi\beta)}{\varphi\beta}\right)\hat{q}_t+\frac{1}{\beta}\hat{q}_{t-1}=-\frac{1}{\beta}z_t. \tag{2.27}$$

Using standard techniques to solve (2.27), yields

$$\hat{q}_t=a\hat{q}_{t-1}+az_t,$$

where a is the root of absolute value less than 1 that solves the characteristic equation: $a^2-\left(\frac{1+\beta}{\beta}+\frac{(1-\varphi)(1-\varphi\beta)}{\varphi\beta}\right)a+\frac{1}{\beta}=0$. It is easy to show that this root is

$$a=\varphi. \blacksquare$$

Proposition 2.2 shows that the parameter φ characterizes both the magnitude and the persistence of the real exchange rate response to a relative money growth shock. The higher φ is, the stronger the shock's initial effect and the more persistent the real exchange rate response will be. More importantly, proposition 2.2 states that φ is precisely the first-order autocorrelation coefficient of the real exchange rate[8]. Hence, real exchange rate persistence is always pinned down by the probability of not changing prices, so that there is no scope for endogenous persistence in this model. Therefore, one cannot replicate the persistence found in real exchange rate movements using a short period of price stickiness. This result parallels the so-called persistence problem raised by Chari, Kehoe, and McGrattan (2000) who show that, in a closed-economy set-up, standard DGE models fail to generate output effects of monetary shocks beyond the exogenous length of price contracts, even if contracts are set in a staggered fashion. Chari, Kehoe, and McGrattan conclude that the contract

[8] In fact, it is easy to show that φ is also the first-order autocorrelation coefficient of the inflation differential $(\hat{\pi}_t-\hat{\pi}^*_t)$ and the relative real money stock $(\hat{m}_t-\hat{m}^*_t)$.

multiplier in standard DGE models is essentially 1[9]. The present study shows that the persistence problem extends to open-economy models. By demonstrating that real exchange rate deviations from the steady state decay at a rate equal to the (exogenous) probability of not changing prices, proposition 2.2 implies that the contract multiplier in this model is exactly equal to 1.

2.4 Robustness

In this section, I assess the sensitivity of my results regarding real exchange rate persistence to alternative specifications of the utility function and the money-growth shock process.

2.4.1 The utility function

The virtue of the functional form (2.1) is that it yields a closed-form solution to the model. However, the logarithmic specification (with respect to real money balances) has an important drawback: it implies an interest elasticity of money demand equal to unity. This is obviously a counterfactual feature, as empirical estimates of such elasticity range from 0.05 in Mankiw and Summers (1986) to 0.39 in Chari, Kehoe, and McGrattan (2000). To allow for plausible values of the interest elasticity of money demand, I consider the following instantaneous utility function:

$$u(c_t, l_t, m_t) = v(c_t, l_t) + \frac{\gamma}{1-\eta} m_t^{1-\eta},$$

where the parameter $\eta > 0$ can be interpreted as the inverse of the interest elasticity of money demand. In the general case where η is different from unity, the eigenvalues of the matrix $\mathbf{A}$ cannot be computed analytically and must, therefore, be evaluated numerically. Table 2.1 shows the "stable" eigenvalue of the matrix $\mathbf{A}$ (which is also the first-order autocorrelation of the model's variables) computed for different values of η ranging from 1 to 20 (implying an interest elasticity of money demand ranging from 1 to 0.05). In this experiment, the parameters β and φ are set to 0.99 and 0.75, respectively.

[9] Chari, Kehoe, and McGrattan (2000) define the contract multiplier as the ratio of the half-life of output to one-half the length of exogenous stickiness. In their model, price rigidity is introduced via Taylor-type staggered contracts, thereby implying that prices fully adjust after the duration of price contracts. With Calvo-type rigidity, however, the aggregate price level, never fully adjusts (by construction). An appropriate definition of the contract multiplier in this case (which shall be used in this paper) would be the ratio of the half-life of a given variable to the half-life of price stickiness.

Table 2.1: The stable eigenvalue of **A** for different values of the parameter η ($\beta = 0.99$ and $\varphi = 0.75$)

η	Stable eigenvalue
1	0.75
2	0.7459
3	0.7420
4	0.7383
5	0.7347
6	0.7313
7	0.7281
8	0.7249
9	0.7219
10	0.7190
11	0.7162
12	0.7134
13	0.7108
14	0.7082
15	0.7052
16	0.7032
17	0.7008
18	0.6985
19	0.6962
20	0.6940

Table 2.1 shows that a plausible parameterization of η does not overturn the lack of persistence generated by the model. In fact, the magnitude of the stable root proves to be a decreasing function of η, meaning that the maximum level of real exchange rate persistence is obtained when $\eta = 1$.

2.4.2 The shock process

As shown earlier, with non-autocorrelated money-growth shocks, the first-order autocorrelation of the real exchange rate is explicitly determined and is equal to the stable eigenvalue of the matrix **A**. When those shocks are serially correlated, however, the first-order autocorrelation of the real exchange rate is no longer pinned down by the magnitude of the stable eigenvalue. To assess the sensitivity of the results to the persistence of money-growth shocks, I assume that the rate of money growth, μ_t, follows a first-order autoregressive process given by

$$\ln \mu_t = (1 - \rho) \ln \mu + \rho \ln \mu_{t-1} + \epsilon_t,$$

where ρ is strictly bounded between -1 and 1, and μ and ϵ_t are defined as in (2.20). The first-order autocorrelation coefficient, ρ, is assumed to be the same for both countries. Under this assumption, the relative shock, z_t, follows a first-order autoregressive process given by

$$z_t = \rho z_{t-1} + \epsilon_t - \epsilon_t^*. \tag{2.28}$$

In this case, the deviation of the real exchange rate from its steady-state value evolves according to

$$\hat{q}_t = \varphi \hat{q}_{t-1} + \phi z_t + \omega z_{t-1},$$

where $\phi = \frac{\varphi(1-\beta\rho+\beta^2\rho-\varphi\beta^2\rho)}{(1-\beta\rho)(1-\varphi\beta\rho)}$, and $\omega = -\frac{\varphi\beta\rho}{1-\beta\rho}$.

Proof.
From equation (2.21), I obtain

$$E_t\left(\hat{\pi}_{t+1} - \hat{\pi}_{t+1}^*\right) = \hat{m}_t - \hat{m}_t^* - E_t(\hat{m}_{t+1} - \hat{m}_{t+1}^*) + \rho z_t. \tag{2.29}$$

Using this equation to substitute for $E_t\left(\hat{\pi}_{t+1} - \hat{\pi}_{t+1}^*\right)$ in (2.23), gives

$$E_t\hat{q}_{t+1} = \frac{1}{\beta}\hat{q}_t + E_t(\hat{m}_{t+1} - \hat{m}_{t+1}^*) - \frac{1}{\beta}(\hat{m}_t - \hat{m}_t^*) - \rho z_t.$$

The solution to this equation is

$$\begin{aligned}\hat{q}_t &= \hat{m}_t - \hat{m}_t^* + \beta\rho\sum_{i=0}^{\infty}\beta^i z_{t+i} \\ &= \hat{m}_t - \hat{m}_t^* + \frac{\beta\rho}{1-\beta\rho}z_t. \end{aligned}\tag{2.30}$$

Using (2.21), (2.29), and (2.30) to substitute for $\hat{\pi}_t - \hat{\pi}_t^*$ and $E_t\left(\hat{\pi}_{t+1} - \hat{\pi}_{t+1}^*\right)$ in (2.22), yields

$$E_t\hat{q}_{t+1} - \left(\frac{1+\beta}{\beta} + \frac{(1-\varphi)(1-\varphi\beta)}{\varphi\beta}\right)\hat{q}_t + \frac{1}{\beta}\hat{q}_{t-1} =$$
$$-\left(\frac{1-\beta\rho}{\beta} + \frac{\rho(1-\beta\rho+\beta)}{1-\beta\rho}\right)z_t + \frac{\rho}{1-\beta\rho}z_{t-1}.$$

The solution to this equation is

$$\begin{aligned}\hat{q}_t &= \varphi\hat{q}_{t-1} + \varphi\beta\left(\frac{1-\beta\rho}{\beta} + \frac{\rho(1-\beta\rho+\beta)}{1-\beta\rho}\right)\sum_{i=0}^{\infty}(\varphi\beta)^i E_t z_{t+i} \\ &\quad -\varphi\beta\frac{\rho}{1-\beta\rho}\sum_{i=0}^{\infty}(\varphi\beta)^i E_t z_{t-1+i}. \end{aligned}\tag{2.31}$$

Finally, noting that $\sum_{i=0}^{\infty}(\varphi\beta)^i E_t z_{t-1+i} = z_{t-1} + \sum_{i=1}^{\infty}(\varphi\beta)^i E_t z_{t+i}$, and using (2.28), equation (2.31) can be written as

$$\hat{q}_t = \varphi\hat{q}_{t-1} + \frac{\varphi(1-\beta\rho+\beta^2\rho-\varphi\beta^2\rho)}{(1-\beta\rho)(1-\varphi\beta\rho)}z_t - \frac{\varphi\beta\rho}{1-\beta\rho}z_{t-1}. \blacksquare$$

Table 2.2: First-order autocorrelation of the real exchange rate for different values of the parameter ρ ($\beta = 0.99$ and $\varphi = 0.75$)

ρ	Autocorrelation
0	0.75
0.1	0.7505
0.2	0.7510
0.3	0.7515
0.4	0.7521
0.5	0.7528
0.6	0.7535
0.7	0.7541
0.8	0.7545
0.9	0.7540

The parameter φ still characterizes the volatility and the persistence of the real exchange rate, but it is no longer interpreted as a first-order autocorrelation coefficient. Table 2.2 shows the serial correlation of the real exchange rate for different values of the parameter ρ[10]. The parameters β and φ are calibrated as in subsection 2.4.1. The table shows that even when the relative money-growth shock is highly autocorrelated, real exchange rate persistence is virtually unaffected, and therefore the contract multiplier remains essentially 1.

2.5 Conclusion

This paper has constructed a tractable DGE two-country sticky-price model whose dynamics can be characterized analytically. Real exchange rate fluctuations generated by the model do not last beyond the average duration of price contracts. More precisely, the contract multiplier delivered by the model is shown to be exactly equal to 1. This outcome is reminiscent of the persistence problem found by Chari, Kehoe, and McGrattan (2000) in a closed-economy set-up.

The failure of sticky-price models to generate persistent real exchange rates suggests that price rigidities might not be the only source driving real exchange rate movements. A similar conclusion is reached by Ng (2003) who uses a VAR-based approach to empirically assess the importance of sticky prices in explaining real exchange rate fluctuations. Her results show that U.S. sticky-price shocks have short half-lives and cannot account for the observed real exchange rate persistence, despite the fact that they explain most of real exchange rate variability.

In view of these results, future research on real exchange rate determination should focus on developing models where real shocks, such as technology and fiscal shocks, affect the real exchange rate in non-trivial ways. Such shocks are

[10] Because money-growth shocks are usually found to be positively autocorrelated in the data, I only focus on positive values of ρ.

likely to be highly persistent and thus, are plausible candidates to account for real exchange rate persistence.

Appendix A: Derivation of equations (2.21)–(2.23)

Derivation of equation (2.21)

Equation (2.19) is approximated as

$$\hat{m}_t = \hat{m}_{t-1} - \hat{\pi}_t + \hat{\mu}_t. \tag{A.2}$$

Taking the difference between (A.2) and its foreign counterpart results in

$$\hat{m}_t - \hat{m}_t^* = \hat{m}_{t-1} - \hat{m}_{t-1}^* - (\hat{\pi}_t - \hat{\pi}_t^*) + \hat{\mu}_t - \hat{\mu}_t^*,$$

which is equation (2.21) in the main text.

Derivation of equation (2.22)

Dividing both sides of equation (2.14) by P_t and using the fact that $MC_\tau/P_t = \left(\Pi_{k=t+1}^{\tau}\pi_k\right) mc_\tau$, I obtain

$$\tilde{p}_{ht} = \frac{\theta}{\theta-1}\frac{E_t\sum_{\tau=t}^{\infty}(\varphi\beta)^{\tau-t}\Lambda_{t,\tau}\left(\Pi_{k=t+1}^{\tau}\pi_k\right) mc_\tau \tilde{y}_{h\tau}}{E_t\sum_{\tau=t}^{\infty}(\varphi\beta)^{\tau-t}\Lambda_{t,\tau}\tilde{y}_{h\tau}},$$

where $\tilde{p}_{ht} = \tilde{P}_{ht}/P_t$. This equation can be approximated as

$$\hat{\tilde{p}}_{ht} = (1-\varphi\beta)\, E_t\sum_{\tau=t}^{\infty}(\varphi\beta)^{\tau-t}\left(\hat{mc}_\tau + \sum\nolimits_{k=t+1}^{\tau}\hat{\pi}_k\right),$$

which can be rewritten in the following recursive form:

$$\hat{\tilde{p}}_{ht} - \varphi\beta E_t\hat{\tilde{p}}_{ht+1} = (1-\varphi\beta)\hat{mc}_t + \varphi\beta E_t\hat{\pi}_{t+1}. \tag{A.3}$$

Similarly, dividing both sides of equation (2.15) by P_t^* and using the fact that $e_\tau P_t^*/P_t = \left(\Pi_{k=t+1}^{\tau}\pi_k\right)\left(\Pi_{k=t+1}^{\tau}\pi_k^*\right)^{-1} q_\tau$, yields

$$\tilde{p}_{ht}^* = \frac{\theta}{\theta-1}\frac{E_t\sum_{\tau=t}^{\infty}(\varphi\beta)^{\tau-t}\Lambda_{t,\tau}\left(\Pi_{k=t+1}^{\tau}\pi_k\right) mc_\tau \tilde{y}_{h\tau}^*}{E_t\sum_{\tau=t}^{\infty}(\varphi\beta)^{\tau-t}\Lambda_{t,\tau}\left(\Pi_{k=t+1}^{\tau}\pi_k\right)\left(\Pi_{k=t+1}^{\tau}\pi_k^*\right)^{-1} q_\tau\tilde{y}_{h\tau}^*}, \tag{A.4}$$

where $\tilde{p}_{ht}^* = \tilde{P}_{ht}^*/P_t^*$. Following the same steps involved in obtaining equation (A.3), it is easy to show that the approximation of equation (A.4) can be written as

$$\hat{\tilde{p}}_{ht}^* - \varphi\beta E_t\hat{\tilde{p}}_{ht+1}^* = (1-\varphi\beta)\left(\hat{mc}_t - \hat{q}_t\right) + \varphi\beta E_t\hat{\pi}_{t+1}^*. \tag{A.5}$$

By analogy to (A.3) and (A.5), the pricing decisions by the foreign monopolistic firm are approximated by

$$\hat{\tilde{p}}_{ft} - \varphi\beta E_t \hat{\tilde{p}}_{ft+1} = (1-\varphi\beta)\left(\hat{mc}_t^* + \hat{q}_t\right) + \varphi\beta E_t \hat{\pi}_{t+1}, \tag{A.6}$$

and

$$\hat{\tilde{p}}^*_{ft} - \varphi\beta E_t \hat{\tilde{p}}^*_{ft+1} = (1-\varphi\beta)\hat{mc}_t^* + \varphi\beta E_t \hat{\pi}^*_{t+1}, \tag{A.7}$$

where $\tilde{p}_{ft} = \tilde{P}_{ft}/P_t$ and $\tilde{p}^*_{ft} = \tilde{P}^*_{ft}/P^*_t$.
Using equations (2.11), (2.16), and its foreign counterpart for P_{ft}, I obtain

$$P_t^{1-\theta} = (1-\varphi)\sum_{\tau=-\infty}^{t} \varphi^{t-\tau}\left[n\tilde{P}_{h\tau}^{1-\theta} + (1-n)\tilde{P}_{f\tau}^{1-\theta}\right]. \tag{A.8}$$

Dividing both sides of equation (A.8) by $P_t^{1-\theta}$ results in

$$1 = (1-\varphi)\sum_{\tau=-\infty}^{t} \varphi^{t-\tau}\left[n\tilde{p}_{h\tau}^{1-\theta}\left(\Pi_{k=\tau+1}^{t}\pi_k\right)^{\theta-1} + (1-n)\tilde{p}_{f\tau}^{1-\theta}\left(\Pi_{k=\tau+1}^{t}\pi_k\right)^{\theta-1}\right].$$

The linearization of this equation yields

$$0 = (1-\varphi)\sum_{\tau=-\infty}^{t} \varphi^{t-\tau}\left[n\left(\hat{\tilde{p}}_{h\tau} - \sum\nolimits_{k=\tau+1}^{t}\hat{\pi}_k\right) + (1-n)\left(\hat{\tilde{p}}_{f\tau} - \sum\nolimits_{k=\tau+1}^{t}\hat{\pi}_k\right)\right],$$

or

$$\begin{aligned}\hat{\pi}_t \;=\; & (1-\varphi)\sum_{\tau=-\infty}^{t}\varphi^{t-\tau-1}\\ & \times\left[n\left(\hat{\tilde{p}}_{h\tau} - \sum\nolimits_{k=\tau+1}^{t-1}\hat{\pi}_k\right) + (1-n)\left(\hat{\tilde{p}}_{f\tau} - \sum\nolimits_{k=\tau+1}^{t-1}\hat{\pi}_k\right)\right].\end{aligned}$$

Subtracting $\varphi\hat{\pi}_t$ from both sides of this equation gives

$$\hat{\pi}_t = \frac{1-\varphi}{\varphi}\left[n\hat{\tilde{p}}_{ht} + (1-n)\hat{\tilde{p}}_{ft}\right]. \tag{A.9}$$

The foreign counterpart of equation (A.9) is

$$\hat{\pi}^*_t = \frac{1-\varphi}{\varphi}\left[n\hat{\tilde{p}}^*_{ht} + (1-n)\hat{\tilde{p}}^*_{ft}\right]. \tag{A.10}$$

Substituting (A.3) and (A.6) into (A.9) and rearranging, I obtain

$$E_t\hat{\pi}_{t+1} = \frac{1}{\beta}\hat{\pi}_t - \frac{(1-\varphi)(1-\varphi\beta)}{\varphi\beta}\left[n\hat{mc}_t + (1-n)(\hat{mc}_t^* + \hat{q}_t)\right]. \tag{A.11}$$

Similarly, equation (A.10), with (A.5) and (A.7) substituted in for $\hat{\tilde{p}}^*_{ht}$ and $\hat{\tilde{p}}^*_{ft}$, becomes

$$E_t\hat{\pi}^*_{t+1} = \frac{1}{\beta}\hat{\pi}^*_t - \frac{(1-\varphi)(1-\varphi\beta)}{\varphi\beta}\left[n(\hat{mc}_t - \hat{q}_t) + (1-n)\hat{mc}^*_t\right]. \qquad (A.12)$$

Finally, subtracting (A.12) from (A.11) yields

$$E_t\left(\hat{\pi}_{t+1} - \hat{\pi}^*_{t+1}\right) = \frac{1}{\beta}\left(\hat{\pi}_t - \hat{\pi}^*_t\right) - \frac{(1-\varphi)(1-\varphi\beta)}{\varphi\beta}\hat{q}_t,$$

which is equation (2.22) in the main text.

Derivation of equation (2.23)

Linearizing the first-order condition (2.5) and its foreign counterpart (with (2.7) substituted in for $\hat{v}^*_{ct}$ and $E_t\hat{v}^*_{ct+1}$) yields, respectively,

$$\hat{m}_t = \frac{\beta}{1-\beta}(E_t\hat{v}_{ct+1} - E_t\hat{\pi}_{t+1}) - \frac{1}{1-\beta}\hat{v}_{ct}, \qquad (A.13)$$

and

$$\hat{m}^*_t = \frac{\beta}{1-\beta}(E\hat{v}_{ct+1} + E_t\hat{q}_{t+1} - E_t\hat{\pi}^*_{t+1}) - \frac{1}{1-\beta}(\hat{v}_{ct} + \hat{q}_t). \qquad (A.14)$$

Subtracting (A.13) from (A.14) and rearranging, I obtain

$$E_t\hat{q}_{t+1} = \frac{1}{\beta}\hat{q}_t - E_t\left(\hat{\pi}_{t+1} - \hat{\pi}^*_{t+1}\right) - \frac{1-\beta}{\beta}\left(\hat{m}_t - \hat{m}^*_t\right),$$

which is equation (2.23) in the main text.

References

Bergin, P. and Feenstra, R. (2001), "Pricing to market, staggered contracts and real exchange rate persistence", *Journal of International Economics*, 54, 333–59.

Betts, C. and Devereux, M. (1996), "The exchange rate in a model of pricing-to-market", *European Economic Review*, 40, 1007–21.

Betts, C. and Devereux, M. (2000), "Exchange rate dynamics in a model of pricing-to-market", *Journal of International Economics*, 50, 215–44.

Blanchard, O.J. and Kahn, C.M. (1980), "The solution of linear difference models under rational expectations", *Econometrica*, 48, 1305–11.

Calvo, G. (1983), "Staggered prices in a utility maximization framework", *Journal of Monetary Economics*, 12, 983–98.

Chari, V., Kehoe, P. and McGrattan, E. (2000), "Sticky price models of the business cycle: Can the contract multiplier solve the persistence problem?", *Econometrica,* 68, 1151–79.

Chari, V., Kehoe, P. and McGrattan, E. (2002), "Can sticky price models generate volatile and persistent real exchange rates ?", *Review of Economic Studies,* 69, 533–63.

Engel, C. (1993), "Real exchange rate and relative prices: An empirical investigation", *Journal of Monetary Economics,* 32, 35–50.

Engel, C. and Rogers, J. (1996), "How wide is the border?", *American Economic Review,* 86, 1112–25.

Knetter, M. (1989), "Price discrimination by U.S. and German exporters", *American Economic Review,* 79, 198–210.

Knetter, M. (1993), "International comparison of pricing-to-market behavior", *American Economic Review,* 83, 473–86.

Kollmann, R. (2001), "The exchange rate in a dynamic optimizing current account model with nominal rigidities: A quantitative investigation", *Journal of International Economics,* 55, 243–62.

Mankiw, N.G. and Summers, L.H. (1986), "Money demand and the effects of fiscal policies", *Journal of Money, Credit and Banking,* 18, 415–29.

Ng, S. (2003), "Can sticky prices account for the variations and the persistence in real exchange rates?", *Journal of International Money and Finance,* Forthcoming.

Obstfeld, M. and Rogoff, K. (1995), "Exchange rate dynamics redux" *Journal of Political Economy,* 103, 624–60.

Part II

Structural breaks and regime switching

Chapter 3

Real exchange rate dynamics: An alternative approach

By Yin-Wong CHEUNG[1] and Ulf G. ERLANDSSON[2]

Abstract

This exercise presents a systematic study on the presence of Markov switching dynamics in three dollar-based real exchange rates. A Monte Carlo approach is adopted to circumvent the statistical inference problem inherent to the test of regime-switching behavior. Annual, quarterly and monthly data are considered. While annual and quarterly data yield inconclusive or weak evidence on the presence of Markov switching behavior, monthly data strongly indicate the presence of multiple regimes in real exchange rate data. The results suggest that data frequency can be crucial for determining the number of regimes.

3.1 Introduction

Real exchange rate dynamics have significant implications for both academia and policymakers. For instance, real exchange rate behavior bears some direct implications for the relevancy of purchasing power parity hypothesis, which is a key assumption in international economics. On the other hand, policymakers are interested in the nature of shocks to real exchange rates that affect trade balances.

Over time, the study of real exchange rate dynamics evolves from simply computing the correlation to sophisticated modeling of nonlinear temporal be-

[1] Corresponding Author: Yin-Wong Cheung, Department of Economics SS1, University of California, Santa Cruz, CA 95044, USA. Email: cheung@cats.ucsc.edu.

[2] Ulf G. Erlandsson, Economics Department, Lund University.

havior. The level of technical complexity in these empirical studies traces the development in (time-series) econometrics. Despite a plethora of empirical studies on real exchange rate behavior, there are still unsettled issues on, for example, the persistence of real exchange rates (Cheung and Lai (2000a, 2000b)), the sources of real exchange rate persistence (Cheung, Lai, and Bergman (2003), Engel and Morley (2001)), and the nonlinear real exchange rate dynamics (Obstfeld and Talyor (1997), Taylor, Peel, and Sarno (2001)). The recent empirical studies on real exchange rate dynamics are surveyed and evaluated in, for example, Rogoff (1996) and Sarno and Taylor (2002).

The purpose of the exercise is to investigate whether real exchange rates follow Markov switching dynamics. Engel and Hamilton (1990), for example, advocate the use of a Markov switching model to describe the behavior of nominal exchange rates. These authors found that a two-regime Markov switching model performs well in both in-sample and out-of-sample periods for three exchange rate series. It is well documented that real exchange rate variability is mainly driven by movements in nominal exchange rates. A Markov switching model for nominal exchange rates is, thus, suggestive of a Markov switching model for real exchange rates. The possibility that real exchange rates follow a Markov switching process is recently explored in Bergman and Hansson (2003).

The presence of regime-switching dynamics has important implications for theoretical models of real exchange rate dynamics. For instance, real exchange rate behavior can depend on how far the real exchange rate is from its equilibrium position (Dumas (1992)). Further, if a real exchange rate "inherits" the time-series property of the corresponding nominal exchange rate, then factors that generate regime-switching behavior in nominal data can also induce regime-switching behavior in real exchange rates. These factors include the "Peso problem", the changing importance of chartists and fundamentalists in the foreign exchange market, differences between domestic and foreign monetary and fiscal policies, and the presence of transaction costs and diversity of opinions (De Grauwe and Vansteenkiste (2001), Engel and Hamilton (1990), Vigfusson (1996)). The relevance of these factors depends critically on whether the observed real exchange rates are generated from multiple regimes. Thus, the research on Markov switching dynamics sheds some useful insight on modeling of real exchange rate behavior.

One main feature of the current exercise is the focus on the determination of the number of regimes in the empirical Markov switching model. Most existing empirical studies on Markov regime-switching dynamics either explicitly or implicitly assume that data are drawn from a specific multi-regime Markov switching process and estimate the model parameters accordingly. The absence of a formal test before estimation is probably due to the fact that the test for the number of regimes is a non-standard statistical inference problem. The commonly used test statistics, including the likelihood ratio statistic, do not have their usual asymptotic distributions. However, without formal evidence of the existence of multiple regimes, estimation of Markov switching models may lead to spurious results. For instance, data generated from a Markov switching model exhibit long swings. However, it is also known that realizations of a random walk process resemble observations displaying long swings. Fitting a Markov switching model to a unit root process may generate spurious estimation results and mis-identify long swings due to unit root persistence as regime-switching behavior. Thus, a formal test for the number of regimes is essential for determining the presence of Markov switching dynamics.

Another interesting feature of the exercise is the attempt to find a level field to determine whether a random walk model or a Markov switching model provides a better description of real exchange rate dynamics. A typical procedure will test the null hypothesis of a random walk against a Markov switching alternative. Such a procedure demands strong evidence to overturn the null hypothesis and establish the alternative specification. It is known that the non-rejection of the null hypothesis does not necessarily mean that the null hypothesis is correct and the alternative is false. If a test procedure has low power, then it may not reject the null hypothesis even if it is false. To mitigate the problem, we consider the null hypothesis of a random walk process and, then, the null hypothesis of a Markov switching process. Thus, both a random walk and a Markov switching model have the same opportunity to be treated "favorably" under the null hypothesis. In addition, we tally the power of the test procedures and incorporate the information in making statistical inferences.

In this study, we adopt a Monte Carlo approach to test for the number of regimes in a Markov switching model (Rydén, Teräsvirta, and Asbrink (1998)). In particular, the data-specific empirical distribution of the likelihood ratio statistic is used to evaluate the probability of the presence of regime-switching behavior. While simulation may not offer general conclusions, it circumvents the issues of non-standard statistical inferences inflicting the standard approach to model regime switching and provides some useful sample-specific results. Further, the use of data-specific distributions helps mitigate finite sample biases.

The remainder of the paper is organized as follows. In Section 3.2 we state the random walk and Markov switching models examined in the exercise and conduct some preliminary analyses on the real exchange rate data. Section 3.3 describes the Monte Carlo based test for the number of regimes. Test results from the annual, quarterly, and monthly data set are also reported in this section. Some concluding remarks are offered in Section 3.4.

3.2 Preliminary analyses

The US dollar real exchange rates of Deutsche mark, British pound, and French franc are considered. The corresponding nominal exchange rates are examined by Engel and Hamilton (1990). The sample spans from 1973 to 1997 and contains data at the monthly, quarterly, and annual frequencies. All data are retrieved from the International Monetary Fund data CD.

It is conceived that the frequency of observations plays a role in revealing regime-switching behavior. Specifically, a higher sampling frequency can give better information on the dynamic property. Suppose a real exchange rate switches between two regimes. If the expected duration of the real exchange rate to stay in one regime is less than a year, then the use of annual data to test for Markov switching dynamics is deemed to be fruitless. Even if the expected duration is a year or two, annual samples still may not offer enough observations within and across realized regimes for the test procedure to disentangle the regime-switching behavior from, say, the random walk behavior. Relatively speaking, monthly observations have a better chance to retain and capture regime-switching behavior. Thus, in this exercise, we explore whether sampling frequency offers some useful information about Markov switching dynamics.

In the following section, we determine whether a random walk (with drift) specification or a two-regime Markov switching model offers a better description of real exchange rate dynamics. These two models are given by

$$\Delta s_t = \mu + \varepsilon_t \tag{3.1}$$

and

$$\Delta s_t = \sum_{i=1,2} I\left(S_t = i\right)\left[\mu_i\right] + \varepsilon_t \tag{3.2}$$

where Δ is the first-difference operator, s_t is the log real exchange rate at time t, μ is the drift term, $\varepsilon_t \sim (0, \sigma^2)$ is an error term, $I(.)$ is an indicator function, $\mu_1 \neq \mu_2$ are the drift terms across regimes 1 and 2, and S_t is the state (regime) variable. The state variable assumes the values of 1 or 2 and follows the transition probabilities $P_{ij}, i, j = 1, 2$, $\sum_{j=1,2} P_{ij} = 1$ for $i = 1, 2$, where P_{ij} is the probability of $S_t = j$ given $S_{t-1} = i$.

Equation (3.1) is the random walk specification. While there is no verdict on whether real exchange rates are stationary, it is difficult for high frequency real exchange rate data to reject the nonstationarity null hypothesis[3]. In this exercise, we take the random walk specification as a reasonable approximation of real exchange rate behavior. The Markov switching model is given by equation (3.2). In essence, the model allows the real exchange rate to behave differently across two regimes defined by the state variable S_t. See, for example, Hamilton (1989) for a discussion on the properties and estimation of a Markov regime-switching model. In this study, we employ a fully unconstrained maximum likelihood method to estimate the model[4].

It is commonly conceived that long swings are found in realizations from a random walk process. Thus, before implementing the Monte Carlo test for the

[3] While there are recent studies indicate that real exchange rates are stationary (Cheung and Lai (2000a)), Engel (1999) argues that the stationarity results may be spurious.

[4] We reproduced the estimates reported in the Engel and Hamilton using their approach assuming a Bayesian Prior. These estimates are virtually the same as those from the fully unconstrained maximum likelihood method. Various randomized starting values are used in the optimization process to ensure that a global maximum likelihood, instead of a local one, is obtained. The fully unconstrained maximum likelihood method is discussed in, for example, Hamilton (1994).

number of regimes, we conducted a pilot simulation study to examine what one may observe when a Markov switching model is fitted to a random walk series. Specifically, we fit a two-regime Markov switching model to artificially generated random walk series. It was found that the average estimates of the transition probabilities P_{11} and P_{22} were indicative of the presence of regime-switching dynamics[5]. Obviously, the pilot study does not rule out the possibility of the existence of Markov switching dynamics in real exchange rate data. However, the result underlines the general conception that long swings in random walk data may lead to spurious evidence of regime-switching behavior and highlights the relevancy of a formal test for the number of regimes.

3.3 Testing for the presence of Markov switching dynamics

In general, the null and alternative hypotheses of the test for the presence of Markov switching can be stated as

- $H0$: the number of regimes in the data is N, and
- $H1$: the number of regimes in the data is $N+1$.

For the current study, we mainly consider the $N = 0$ case. That is, the data are drawn from a single regime under the null hypothesis and from two distinct regimes under the alternative. Let θ_N and θ_{N+1} be the parameter vectors under the null and alternative hypotheses and $\hat{\theta}_N$ and $\hat{\theta}_{N+1}$ be the corresponding maximum likelihood estimators. The likelihood ratio statistic is

$$LR = 2\left[L\left(\hat{\theta}_{N+1}\right) - L\left(\hat{\theta}_N\right)\right] \tag{3.3}$$

where $L\left(\hat{\theta}_{N+1}\right)$ and $L\left(\hat{\theta}_N\right)$ are the log likelihood functions evaluated at their respective maximum likelihood estimators $\hat{\theta}_{N+1}$ and $\hat{\theta}_N$.

In testing for Markov switching, the complexity arises because some parameters are not identified under the null. The presence of un-identified parameters invalidates the regularity conditions under which the likelihood ratio statistic can be shown to follow an asymptotic chi-square distribution. Several procedures have been suggested to test for the number of regimes; see, for example, Hansen (1992), Gong and Mariano (1997), and Garcia (1998). These procedures typically derive the asymptotic behavior of the test statistic under some pre-specific conditions and offer different finite-sample performance for different model configurations[6].

[5] Specifically, the averages of the estimated transition probabilities obtained from the simulation exercise are 0.89.

[6] The Hansen (1992) test, for instance, involves a grid search over the parameter space and can be computationally complicated for a multi-regime dynamic model. The Garcia (1998)

Following the lead of Rydén, Teräsvirta, and Asbrink (1998), a Monte Carlo approach is adopted to derive the empirical distribution of the likelihood ratio statistic, which is then used to determine the significance of the statistic computed from the actual data. For a given real exchange rate series, the empirical distribution for testing $H0$ against $H1$ is constructed according to these steps: a) find the maximum likelihood estimator $\hat{\theta}_N$ that gives the best specification within the class of $N-$regime models, b) use $\hat{\theta}_N$ to generate a sample of the artificial data, c) compute the likelihood statistic (3.3) from the generated data, and d) repeat steps (b) and (c) M times and store the simulated likelihood ratio statistics. The M simulated likelihood ratio statistics are used to derive the empirical distribution to test the null. Specifically, the empirical $p-$value of the Markov switching test is given by $(m+1)/(M+1)$, where m is the number of simulated statistics that are larger than the likelihood ratio statistic computed from the actual data series.

3.3.1 Annual data

Random walk against Markov switching

Let us consider the case in which (3.1) specifies the null hypothesis and (3.2) is the Markov switching model under the alternative. The sample likelihood ratio statistics are given in the second column ("Sample LR") of Panel A in table 3.1. The Deutsche mark and French franc series give similar sample likelihood ratio test statistics, which are smaller than the one computed from the and British pound data.

For each exchange rate series, the sample-specific empirical distribution of the likelihood ratio statistic is generated according to the procedure outlined above. Due to the computational complexity of the simulation experiment, the number of replications M is set to 250[7]. Some descriptive statistics of these empirical distributions and the associated $p-$values are given in table 3.1. All the empirical distributions have their means larger than the medians. Further, these are positively skewed distributions; that is, they have a long tail to the right. The ranges of these simulated statistics vary across real exchange rates. In general, these descriptive statistics indicate that the empirical distributions are series-specific.

---

and Gong and Mariano (1997) procedures are easier to implement but not much is known about their small sample performance. It should be noted that these procedures focus on testing between one or two regimes and become quite complicated when a higher dimension model is considered. On the other hand, the Monte Carlo approach adopted in this exercise can be readily modified to handle higher order Markov switching dynamics.

[7] Rydén, Teräsvirta, and Asbrink (1998), for example, set $M = 50$. For 250 replications and a 10% test used in the subsequent simulation, the 95% confidence interval of the rejection frequency is 10% ± 3.7% under the null hypothesis. Thus, even after taking sampling uncertainties into consideration, the inferences presented in the text are not affected.

Table 3.1: Random walk or Markov switching. Annual data

	Sample LR	P-Value	Mean	Median	Std Error	Skew.	Max.	Emp. power
			Panel A: Random Walk against Markov Switching					
DEM	2.418	0.891	5.619	5.153	2.727	0.755	13.010	0.100
GBP	8.263	0.187	5.678	5.152	3.027	0.832	15.710	0.564
FFR	3.996	0.661	5.660	4.848	3.251	0.848	19.671	0.152
			Panel B: Markov Switching against Random Walk					
DEM	2.418	0.030	5.736	5.426	3.105	0.581	15.629	0.170
GBP	8.263	0.291	11.817	11.422	6.087	0.536	32.003	0.288
FFR	3.996	0.299	6.319	5.654	3.391	1.302	24.217	0.136

The likelihood ratio statistics computed from the annual real exchange rate data (DEM = Deutsche mark, GBP = British pound, FFR = French franc) are reported under the column labeled "Sample LR". The p-values of these sample statistics derived from the empirical distributions are listed under "P-Value". Descriptive statistics of the empirical distributions are provided under "Mean", "Median", "Std Error", "Skew." (skewness) and "Maximum".
For Panel A, the empirical distributions of the likelihood ratio statistic are generated from random walk models estimated from the data. The column "Emp. Power" (empirical power) reports the frequency of rejections when these empirical distributions are used to evaluate the significance (at the 10% level) of the likelihood ratio statistics calculated from data generated from Markov switching processes estimated from real exchange rate data.
For Panel B, the empirical distributions of the likelihood ratio statistic are generated from Markov switching models estimated from the data. The column "Empirical Power" reports the frequency of rejections when these empirical distributions are used to evaluate the significance (at the 10% level) of the likelihood ratio statistics calculated from data generated from random walk processes estimated from real exchange rate data. In all cases, the number of replications is 250. The sample period is 1973 to 1997.

The empirical distributions of the likelihood ratio statistic cast doubts on the presence of Markov switching dynamics. According to the $p-$values, there is no strong evidence to reject the random walk null in favor of the Markov switching model. The $p-$values for the Deutsche mark and French franc data are 66% or higher. The one for British pound series is below 20%, which is a rather weak evidence against the random walk hypothesis. Thus far, the evidence is in favor of the random walk null hypothesis.

Are the non-rejection results driven by the low power of the testing procedure? To address the question on power, we generate artificial data series from two-regime Markov switching models estimated from individual real exchange rate series and calculate the likelihood ratio statistic. A 10% test (with the critical value from the corresponding simulated null distribution) is used to determine the proportion of rejection, which is defined as the empirical power. The empirical power is reported in the last column of table 3.1. The power of the test against the Deutsche mark and French franc Markov switching specifications is quite low — 10% for the Deutsche mark case and 15% for the French franc case. Nonetheless, more than 56% of the simulated British pound Markov switching series are rejected at the 10% level. While power can be an explanation of the non-rejection result for the Deutsche mark and French franc cases, it seems not relevant for the British pound result.

Markov switching against random walk

One potential criticism of the results reported in the previous subsection is that the inferences are made with the random walk specification as the null hypothesis. The setting gives the random walk specification the benefit of the doubt. Specifically, the random walk specification is rejected only if there is strong evidence against it. In this subsection, we consider a heuristic approach to account for this asymmetric treatment.

A natural question to ask is "What is the behavior of the likelihood ratio statistic if the data are in fact generated from a Markov switching model?" If the Markov switching model gives the true data generating process, we expect there is substantial gain in the likelihood value between $L\left(\hat{\theta}_{N+1}\right)$ and $L\left(\hat{\theta}_{N}\right)$ and, hence, the likelihood ratio statistic to be large. Alternatively, if a random walk model generates the data, the likelihood value will tend to be small. Heuristically speaking, when the likelihood ratio statistic computed from the actual data is small compared with the values of the likelihood ratio statistic generated from simulated Markov switching data, it can be regarded as evidence against the Markov switching model and in favor of the random walk specification. Thus, we simulate the data under a Markov switching specification and tally the likelihood ratio statistics from these artificial data series. Then, we look at whether the likelihood ratio statistic computed from exchange rate data is "significantly" smaller than the simulated likelihood ratios. The notion of the presence of Markov switching dynamics is rejected only if there is strong evidence against it — that is, the Markov switching model is rejected if there is only a very small portion of the simulated statistics is less than the likelihood ratio statistic from the actual data. In doing so, we reverse the asymmetric treatment built into the procedure considered in the previous subsection.

Panel B of table 3.1 reports the results of testing for Markov switching dynamics against random walk alternatives. Artificial annual data series are generated according to Markov switching models fitted to the three real exchange rate series. The likelihood ratio statistic (3.3) is constructed from each replication and tabulated to derive the empirical distribution to determine the significance of the sample likelihood ratio statistic. The "$p-$value" is calculated as $m/(M+1)$, where m is the number of simulated statistics that are smaller than the likelihood ratio statistic computed from actual data. Compared with Panel A, the descriptive statistics in Panel B are consistent with the prior belief that the likelihood ratio statistic under the Markov switching model tends to be large. Again, these empirical distributions are positively skewed and data-specific.

For the British pound and French franc series, the results show that the sample likelihood ratio statistics are not too small compared with the simulated values. In these two cases, the sample likelihood ratio statistics are larger than at least 29% of the statistics generated from simulated Markov switching data. The results do not support the notion that the sample statistics are substantially less than what would be expected if data were generated via the Markov switching channel. Alternatively, we can say that, for these two real exchange rate series, there is no strong evidence to reject the Markov switching model in favor of the random walk model.

On the other hand, the simulation result indicates the rejection of the Markov switching model for the Deutsche mark series. The sample likelihood statistic computed from the Deutsche mark data is less than 97% of the statistic simulated from a Markov switching specification. The power of the test, which is listed in the last column of Panel B, is in the range of 14% to 29%. That is, if the data are simulated with the random walk specifications, then 14% to 29% of the simulated series give a likelihood ratio statistic that is smaller than the 10th percentile of the empirical distributions based on the Markov switching models. Given the low power of the procedure, it is not sure whether the rejection in the Deutsche mark case is a statistical artifice or is a genuine piece of evidence that the Deutsche mark real exchange rate follows a random walk.

The non-rejection of Markov switching result is enlightening. It underscores a fundamental issue in hypothesis testing — the non-rejection of the null hypothesis does not necessarily mean the null hypothesis is correct. In fact, what the results in table 3.1 indicate that, for the British pound and French franc real exchange rates and the parametric information retrieved from the data, we reject neither the random walk nor the Markov switching specification. In the following, we will examine the robustness of the Deutsche mark random walk result.

As attested by the simulation results, the Monte Carlo based tests may not offer a definite evaluation on whether there is Markov switching dynamics in the annual data. The inference depends on which specification is being considered as the null hypothesis. Seemingly, annual data may not contain sufficient information to discriminate between the two alternatives.

3.3.2 Quarterly and monthly data

To explore the implication of data frequency, we examine quarterly and monthly data of the same three real exchange rates. The results derived from quarterly and monthly data are summarized in table 3.2 and table 3.3, respectively. Again, for both data frequencies, we a) test the null hypothesis of a random walk against the alternative of a Markov switching model and report the results in Panel A of the tables, and b) test the Markov switching null against the random walk alternative and report the results in Panel B of the tables. Artificial data generated according to random walk and Markov switching specifications estimated from real exchange rates are used to compute data-specific size and power estimates. For both quarterly and monthly data, the empirical distributions constructed under a random walk model and a Markov switching model are sample-specific, and, thus, reinforce the importance of using data-specific information to evaluate the relative performance of the competing models.

Compared with annual data, the quarterly data offer stronger evidence of Markov switching behavior (table 3.2). The random walk model is strongly rejected with a $p-$value of 0.4% in the French franc case, rejected with a $p-$value of 2.8% in the Deutsche mark case, and not rejected in the British pound case.

The empirical power of the test is higher than that reported for the annual data. Indeed, the empirical power in Panel A of table 3.2 ranges from 53% to 94%. Very likely, the power improvement attributes to the rejection results associated with the quarterly data.

Table 3.2: Random walk or Markov switching. Quarterly data

	Sample LR	P-Value	Mean	Median	Std Error	Skew.	Max.	Emp. power
			Panel A: Random Walk against Markov Switching					
DEM	14.560	0.028	7.123	6.326	2.799	1.414	19.077	0.724
GBP	9.344	0.211	7.286	6.637	2.972	1.350	20.582	0.532
FFR	17.887	0.004	6.571	5.866	2.499	1.377	17.218	0.936
			Panel B: Markov Switching against Random Walk					
DEM	14.560	0.502	15.685	14.525	6.217	0.661	34.641	0.772
GBP	9.344	0.299	12.035	11.119	4.440	0.901	25.716	0.580
FFR	17.887	0.514	19.069	17.699	7.327	1.305	60.667	0.940

The likelihood ratio statistics computed from the quarterly real exchange rate data (DEM = Deutsche mark, GBP = British pound, FFR = French franc) are reported under the column labeled "Sample LR". The sample period is 1973:I to 1997:IV. See the note to table 3.1.

Table 3.3: Random walk or Markov switching. Monthly data

	Sample LR	P-Value	Mean	Median	Std Error	Skew.	Max.	Emp. power
			Panel A: Random Walk against Markov Switching					
DEM	26.191	0.004	6.402	5.551	2.882	1.467	17.324	0.968
GBP	16.308	0.004	6.296	5.714	2.588	1.049	15.910	0.928
FFR	26.708	0.004	6.200	5.492	2.795	1.592	18.653	0.968
			Panel B: Markov Switching against Random Walk					
DEM	26.191	0.681	23.510	22.610	8.932	0.947	56.961	0.960
GBP	16.308	0.426	19.047	17.691	7.760	0.861	48.448	0.920
FFR	26.708	0.634	24.303	23.450	9.221	0.571	49.170	0.972

The likelihood ratio statistics computed from the monthly real exchange rate data (DEM = Deutsche mark, GBP = British pound, FFR = French franc) are reported under the column labeled "Sample LR". The sample period is 1973:1 to 1997:12. See the note to table 3.1.

The results in Panel B of table 3.2 do not represent any significant evidence against the Markov switching model. The lack of power does not appear to be a relevant explanation of the non-rejection results. The ability of the procedure to reject a random walk model is pretty high. The simulated empirical power is in the range of 58% to 94%, which is comparable to the power figures in Panel A. One interesting observation is that the annual and quarterly Deutsche mark real exchange rate series give different results from the test of the Markov switching model against the random walk model — the former series rejects the Markov switching specification while the latter does not. Comparing the empirical power in tables 3.1 and 3.2, we can rule out the possibility that the quarterly data non-rejection result is due to the power of the test. Thus, it is possible that the annual frequency result is a statistical artifice because the null hypothesis, even though it is correct, can be rejected by chances. We will get back to this point again in analyzing monthly data.

Table 3.3 reports the results obtained from monthly data. Contrary to annual and quarterly data, the monthly data yield a much sharper inference on real exchange rate dynamics. For all the three monthly real exchange rate series, the Monte Carlo based test rejects the random walk null hypothesis with a $p-$value 0.004. When Markov switching is assumed to be the data generating process, the sample likelihood ratio statistics are larger than 43% to 68% of the simulated statistics — indicating that the sample statistics are not likely to come from a random walk process. Further, the procedures have very good power against their respective alternatives. In all cases, the simulated power is larger than 90%. Regarding the Deutsche mark case, the result from the monthly data is similar to the one from quarterly data but different from annual data. Taking all the evidence into consideration, it is reasonable to infer that the Deutsche mark data have switching dynamics and the annual data result is likely to be spurious. Overall, there is strong evidence of monthly real exchange rates following Markov switching dynamics.

One observation comes out pretty clearly in the annual, quarterly and monthly results. An increase in sampling frequency has an appreciable effect on the test performance. Specifically, high frequency data allow the observations to display switching behavior that may have duration too short to be observed in low frequency data. Further, an increase in sampling frequency, for a given sample time span, increases the number of observations and boosts the power of the test procedure. These two factors together offer a better chance to detect Markov switching dynamics in data.

3.4 Concluding remarks

The study presents a systematic empirical examination on the presence of Markov switching dynamics in three dollar-based real exchange rate series. Annual, quarterly, and monthly data from 1973 to 1997 are considered. A Monte Carlo approach is adopted to circumvent the statistical inference problem inherent to the modeling of regime switching. To address the asymmetry of having the random walk as the null hypothesis, we also examine the empirical behavior of the likelihood ratio statistic under the sample-specific Markov switching models. Because the behavior of the statistic is found to be contingent on the sample-specific dynamics, sample-specific empirical distributions of the likelihood statistic are used to infer the relative performance of the random walk and Markov switching models.

The ability to discriminate between random walk and Markov switching depends on the power of the test and the information content of the data. The current exercise illustrates that the Monte Carlo based test can be a promising procedure to detect Markov switching dynamics. The results obtained from three data frequencies are consistent with the informational interpretation. Better information about Markov switching dynamics can be obtained by increasing the sampling frequency. Our results buttress the importance of using high frequency data to extricate Markov switching dynamics.

References

Bergman, U.M. and Hansson, J. (2003), "Real exchange rates and switching regimes", forthcoming, *Journal of International Money and Finance.*

Cheung, Y.W. and Lai, K.S. (2000a), "On cross-country differences in the persistence of real exchange rates", *Journal of International Economics*, 50, 375-397.

Cheung, Y.W. and Lai, K.S. (2000b), "On the Purchasing Power Parity puzzle", *Journal of International Economics*, 52, 321-330.

Cheung, Y.W., Lai, K.S. and Bergman, M. (2003), "Dissecting the PPP puzzle: The unconventional roles of nominal exchange rate and price adjustments", forthcoming, *Journal of International Economics.*

De Grauwe, P. and Vansteenkiste, I. (2001), "Exchange rates and fundamentals: A non-linear relationship?", *Working Paper*, CEPR and University of Leuven.

Dumas, B. (1992), "Dynamic equilibrium and the real exchange rate in a spatially separated world", *Review of Financial Studies*, 5, 153-180.

Engel, C. (1999), "Long-run PPP may not hold after all", *Journal of International Economics*, 51, 243-273.

Engel, C. and Hamilton, J. D. (1990), "Long swings in the Dollar: Are they in the data and do markets know it?", *American Economic Review*, 80, 689–713.

Engel, C. and Morley, J.C. (2001), "The adjustment of prices and the adjustment of the exchange rate", *NBER Working Paper* #8550.

Garcia, R. (1998), "Asymptotic null distribution of the likelihood ratio test in Markov switching models", *International Economic Review*, 39, 763–788.

Gong, F. and Mariano, R.S. (1997), "Testing under non-standard conditions in frequency domain: With applications to Markov regime switching models of exchange rates and the Federal funds rate", Federal Reserve Bank of New York Staff Reports, No. 23.

Hamilton, J.D. (1989), "A new approach to the economic analysis of nonstationary time series and the business cycle", *Econometrica*, 57, 357-84.

Hamilton, J.D. (1994), *Time series analysis*, Princeton University Press.

Hansen, B.E. (1992), "The likelihood ratio test under non-standard conditions: Testing the Markov switching model of GNP", *Journal of Applied Econometrics*, 7, S61-S82.

Obstfeld, M. and Taylor, A.M. (1997), "Nonlinear aspects of goods-market arbitrage and adjustment: Heckscher's commodity points revisited", *Journal of Japanese and International Economics*, 11, 441-479.

Rogoff, K. (1996), "The Purchasing Power Parity puzzle", *Journal of Economic Literature*, 34, 647-668.

Sarno, L. and Taylor, M.P. (2002), "Purchasing Power Parity and the real exchange rate", *IMF Staff Papers*, 49, 65-105.

Taylor, M.P., Peel, D.A. and Sarno, L. (2001), "Nonlinear mean-reversion in real exchange rates: Towards a solution to the Purchasing Power Parity puzzles", *International Economic Review*, 42, 1015-1042.

Rydén, T., Teräsvirta, T. and Asbrink, S. (1998), "Stylized facts of daily return series and the hidden Markov model", *Journal of Applied Econometrics*, 13, 217–244.

Vigfusson, R. (1996), "Switching between chartists and fundamentalists: A Markov regime-switching approach", *Bank of Canada Working Paper* 96-1.

Chapter 4

Structural breaks and unit roots in black market real exchange rates

By Mario CERRATO[1] and Nicholas SARANTIS[2]

Abstract

We use a unique set of data on monthly black market exchange rates from twenty emerging market economies to investigate the presence of structural breaks and unit roots in real exchange rates. Our findings support the hypothesis that real exchange rates are break-trend stationary in nine emerging markets, but fail to find evidence of mean reversion in the other countries.

4.1 Introduction

Long-run Purchasing Power Parity (PPP) is a major assumption in most theoretical models in open economy macroeconomics. Evidence on persistent deviations from PPP would undermine these models and have important policy implications. Most of the empirical literature on PPP is based on testing for unit roots in the real exchange rate and focuses primarily on the experience of industrial countries, with mixed results (see Sarno and Taylor (2002)). Very little work has been done for emerging market economies, especially using black

[1] Corresponding author: Mario Cerrato, Department of Economics, Finance & International Business, London Metropolitan University, 31 Jewry Street, London EC3N 2EY, UK. Tel.: 020 7320 1382. Email: m.cerrato@londonmet.ac.uk.

[2] Department of Economics, Finance and International Business, London Metropolitan University, UK.

market exchange rates[3], which play such a major role in those economies. The only exceptions are Phylaktis and Kosimmatis (1994) and Luintel (2000), but both these studies cover only a small number of countries.

Emerging market economies have experienced different exchange rate regimes and major policy shifts and major devaluations over the last thirty years, which raises the possibility of structural shifts in the underlying process of the real exchange rate. It is therefore essential that any application of unit root tests to the real exchange rates of emerging markets allows for structural breaks[4].

This paper contributes to the PPP literature by using a new data set of monthly black market exchange rates for twenty emerging market economies up to the 1990s. To our knowledge, empirical investigation of real exchange rates using black market rates of this dimension has not been previously undertaken. We also contribute to the debate on structural breaks in unit root tests of the PPP, where there has been very little work.

The paper is organized as follows. Section 4.2 outlines the tests and estimation procedures. Section 4.3 discusses the data, while the empirical results are presented in Section 4.4. The final section provides a summary of the findings.

4.2 Sequential tests of the unit root and trend-break stationary hypotheses

We apply the methodologies of Banerjee *et al.* (1992) and Zivot and Andrews (1992) which allow us to estimate the break date endogenously by making it data dependent, and then to test for the unit root hypothesis conditional on the identified structural break. It is assumed that the null hypothesis is a unit root process without a structural break[5], while the alternative hypothesis is a trend-stationary process with a structural break. Consider the following models for the real exchange rate (q) which nest the null and the appropriate alternative hypotheses:

$$q_t = \alpha^A + \theta^A DU_t(\lambda) + \beta^A t + \delta^A q_{t-1} + \sum_{i=1}^{k} c_i \Delta q_{t-i} + e_t \tag{4.1}$$

[3]Black market exchange rates have a long tradition in emerging market economies. In fact, in many cases the volume of transactions in these markets was much larger than that in the official market.

[4]The issue of structural breaks in the real exchange rate is also raised by Papell (2002), though his investigation is restricted to industrial countries. The only study that tests for structural breaks in black market real exchange rates is Luintel (2000), but his analysis considers eight Asian countries only and covers the period up to the 1980s.

[5]Vogelsang and Perron (1998) show that for structural breaks typically encountered in practice, the finite sample distributions are adequately approximated by the asymptotic distributions of these tests that are obtained under the assumption of no break under the unit root null hypothesis.

$$q_t = \alpha^B + \beta^B t + \eta^B DT_t^* (\lambda) + \delta^B q_{t-1} + \sum_{i=1}^{k} c_i \Delta q_{t-i} + e_t \qquad (4.2)$$

$$q_t = \alpha^C + \theta^C DU_t (\lambda) + \beta^C t + \eta^C DT_t^* (\lambda) + \delta^C q_{t-1} + \sum_{i=1}^{k} c_i \Delta q_{t-i} + e_t \quad (4.3)$$

where λ denotes the break date with $1 < \lambda < T$; T is the sample size; $DU_t(\lambda) = 1$ if $t > \lambda$, and 0 otherwise; $DT_t^* (\lambda) = t - \lambda$ if $t > \lambda$, and 0 otherwise.

Model (4.1) is referred to as the crash model that allows one time shift in the intercept of the trend, while model (4.2) is referred to as the changing growth model that allows a shift in the slope of the trend function (see Perron (1989)). Model (4.3) allows simultaneously for a shift in both the intercept and slope of the trend function (mixed model). Banerjee *et al.* (1992) considered only the first two models, while Zivot and Andrews (1992) examined all three models.

The true break date, λ^c, is assumed to be in the interval $\Psi = \{\lambda_0, \lambda_0 + 1, ..., T - \lambda_0\}$, where $\lambda_0 = \gamma_0 T$ is the initial start up sample and γ_0 is the trimming parameter. Equations (4.1)-(4.3) are estimated for the break dates $\{\lambda_0, \lambda_0 + 1, ..., T - \lambda_0\}$ and the sequence of ADF statistics for $H0 : \delta = 1$, denoted as $t_{ADF^i} (T_b)$ for $T_b = [\lambda_0 \text{ up to } T - \lambda_0]$ with $i = 1, 2, 3$, are computed. The minimum sequential ADF statistic, as suggested in Banerjee *et al.* (1992) and Zivot and Andrews (1992), is given by the statistic that maximizes the evidence against the no-break unit root null hypothesis.

4.3 The data

We use monthly data on the black market exchange rates for a panel of twenty emerging market countries (Nepal, Pakistan, Philippines, S. Lanka, Thailand, Turkey, Venezuela, Indonesia, Kenya, Korea, Malaysia, Ethiopia, Ghana, Hungary, India, Algeria, Bolivia, Colombia, D. Republic, Egypt) over the period 1973M1-1993M12. The sample ends in 1993 because of the unavailability of data beyond that year. The US Dollar is used as numeraire currency. The black market exchange rates are obtained from *Pick's Currency Yearbook* (various publications), and from the *World Currency Yearbook* (various issues) published by the International Currency Analysis. The consumer price index (CPI) is used as price index. The real exchange rate is given by $q_t = s_t - p_t - p_t^*$, where s is the nominal exchange rate (domestic currency price of the US dollar), p is the domestic price and p^* is the foreign (USA) price, all measured in logarithms.

The black market real exchange rates for the twenty emerging market economies are displayed in the appendix. These graphs appear to indicate segmented trends for some countries. The upward trend in the real exchange rate for most countries seems to suggest that many emerging market economies experienced real depreciation during our sample period.

Table 4.1: Sequential tests of structural breaks

Country	Mean-shift (4.1) t-min(ADF)	Trend-shift (4.2) t-min(ADF)	Mean and trend shift (4.3) t-min(ADF)
Algeria	-2.12 [82:04]	-2.78 [80:08]	-3.11 [76:01]
Bolivia	-14.4*** [86:11]	-2.46 [82:05]	-13.84*** [86:11]
Colombia	-2.39 [85:01]	-2.02 [75:12]	-2.31 [75:12]
Dom. Rep.	-3.42 [85:01]	-3.67 [83:07]	-3.81 [83:11]
Egypt	-3.09 [85:01]	-3.87 [80:12]	-3.71 [80:12]
Ethiopia	-3.99 [78:02]	-4 [80:12]	-4.23 [75:12]
Ghana	-7.21*** [75:12]	-4.81** [78:02]	-7.52*** [75:12]
Hungary	-3.91 [89:12]	-3.29 [86:01]	-4.03 [82:05]
India	-4.12 [86:11]	-3.98 [89:12]	-4.58 [86:01]
Indonesia	-2.26 [75:12]	-1.57 [78:02]	-2.65 [78:02]
Kenya	-3.16 [89:12]	-3.64 [89:12]	-3.71 [89:12]
Korea	-4.6* [82:05]	-4.67** [87:07]	-5.02* [85:01]
Malaysia	-4.27 [86:01]	-4.34* [89:12]	-4.52 [86:01]
Nepal	-4.62* [85:01]	-4.76** [75:12]	-4.93* [86:01]
Pakistan	-4.44 [75:12]	-4.90** [78:02]	-4.97* [85:01]
Philippines	-4.81** [82:05]	-3.81 [89:12]	-4.63 [82:05]
S. Lanka	-4.75* [78:02]	-5.38*** [78:02]	-5.39** [78:02]
Thailand	-4.63* [83:11]	-4.23* [89:12]	-4.2 [86:11]
Turkey	-2.67 [80:12]	-3.38 [83:11]	-3.51 [82:05]
Venezuela	-3.84 [82:05]	-2.63 [87:07]	-3.73 [82:05]
CV 1%	-5.34	-4.93	-5.57
CV 5%	-4.80	-4.42	-5.08
CV 10%	-4.58	-4.11	-4.82

The critical values (CV) for min t_{ADF} are taken from Zivot and Andrews (1992), Tables 2-4. The symbols ***, **, and * indicate significance at the 1%, 5% and 10% levels, respectively. Numbers within brackets denote the date of the break chosen by the sequential procedure (i.e. the one that corresponds to the minimum t_{ADF} value).

4.4 Empirical results

Table 4.1 presents the sequential ${t_{ADF}}^i$ statistics for models (4.1)-(4.3). The lag length used in the ADF framework, k, is selected with the Campbell and Perron (1991) procedure. Critical values for these statistics are reported at the bottom of table. The trimming parameter γ_0 is set to 0.15, as suggested in Banerjee *et al.* (1992).

Let us consider the mean shift model first. We can strongly reject the null hypothesis of a unit root only in three countries, namely Ghana, Bolivia and Philippines. There is weaker evidence (at the 10% level) against the unit root null for Korea, Nepal, S. Lanka and Thailand. In the case of the trend shift model, the unit root null hypothesis is rejected in five countries (*i.e.* Ghana, Korea, Nepal, Pakistan and S. Lanka) at the 5% or 1% levels, and in two countries (*i.e.* Malaysia and Thailand) at the lower significance level of 10%. Finally, the results for the mixed (mean and trend shift) model suggest rejection

Table 4.2: Estimates of the mean-shift model (4.1)

Country	T_b	α	θ	β	δ	$S(e)$
Bolivia	1986:11	7.064 [13.86]	−6.861 [−14.37]	0.003 [2.40]	0.527 [−14.42]	0.657
Philippines	1982:05	0.512 [4.80]	0.053 [3.34]	$-7E^{-05}$ [−0.79]	0.842 [−4.81]	0.052
Thailand	1983:11	0.684 [4.61]	0.031 [2.81]	0.0001 [1.89]	0.778 [−4.63]	0.034

Statistics in brackets are $t-$values. The $t-$statistic for δ is for testing the hypothesis $\delta = 1$ (CV: -5.34(1%), -4.80(5%), -4.582(10%)). $S(e)$ is the standard error of the regression.

Table 4.3: Estimates of the trend-shift model (4.2)

Country	T_b	α	β	η	δ	$S(e)$
Korea	1987:07	2.493 [4.65]	0.0004 [2.17]	−0.002 [−2.63]	0.630 [−4.67]	0.109
Malaysia	1989:12	0.063 [4.19]	0.0004 [4.40]	−0.0004 [−2.42]	0.867 [−4.34]	0.023
Nepal	1975:12	0.484 [4.44]	0.002 [2.44]	−0.002 [−1.78]	0.830 [−4.76]	0.075
Pakistan	1978:02	0.747 [4.78]	0.002 [4.42]	−0.001 [−2.94]	0.726 [−4.90]	0.054
S. Lanka	1978:02	1.957 [5.23]	0.008 [3.80]	−0.008 [−3.48]	0.399 [−5.38]	0.314

Statistics in brackets are $t-$values. The $t-$statistic for δ is for testing the hypothesis $\delta = 1$ (CV: -4.93(1%), -4.42(5%), -4.11(10%)). $S(e)$ is the standard error of the regression.

of the unit root null at the 1% level for Ghana and Bolivia, at the 5% level for S. Lanka, and at the 10% for Korea, Nepal and Pakistan[6]. It seems that Ghana, Korea, Nepal and S. Lanka are the most frequently selected countries. Out of the nine countries where the unit root null is rejected, the weakest evidence is for Malaysia, Thailand and Philippines.

In cases where the unit root null hypothesis is rejected, we can determine the significance of the other coefficients given that their $t-$values are normally distributed (see Perron (1989), Sen (2000)). For countries where the unit root null is rejected for more than one specification of the break, we report estimates of the model with the highest significance level of the null hypothesis. If the highest significance level is the same for two models, then we report estimates of the model with the lowest t_{ADF} statistic.

The results for the mean-shift model (4.1) are shown in table 4.2. The constant and the mean shift dummy are highly significant in all three countries, while the trend is significant in two countries (Bolivia and Thailand). The trend-shift model (4.2) is chosen for most countries that experienced a structural break and its estimates are reported in table 4.3. It is interesting to note that the constant, the trend and the dummy for the shift in the slope of the trend are highly significant in all five countries, with the trend displaying a downwards

[6] Our results contrast with Papell's (2002) evidence against break-trend stationarity in the real exchange rate for all individual industrial countries (except France), and with Luintel's (2000) findings for the eight Asian countries (except for Malaysia and Thailand).

Table 4.4: Estimates of the mean-shift and trend-shift model (4.3)

Country	T_b	α	θ	β	η	δ	$S(e)$
Ghana	1975:12	2.001	0.483	−0.007	0.007	0.681	0.163
		[7.21]	[5.72]	[−2.00]	[1.97]	[−7.52]	

Statistics in brackets are t−values. The t−statistic for δ is for testing the hypothesis $\delta = 1$ (CV: -5.57(1%), -5.08(5%), -4.82(10%)). $S(e)$ is the standard error of the regression.

shift. The mixed model is chosen only for Ghana and the estimates are displayed in table 4.4. Again all coefficients are statistically significant. An interesting observation is that the trend has a positive coefficient in seven out of the nine countries where there is evidence of trend-break stationarity, thus implying rising real appreciation over time. The break dates differ across countries, which suggests that they are likely to be country-specific.

4.5 Conclusions

This paper reports evidence on structural breaks and mean reversion in real exchange rates for twenty emerging market economies, using a new set of data for black market exchange rates. We find considerable evidence of structural breaks, which are determined endogenously. In nine out of twenty countries, the black market real exchange rate is generated by a trend-stationary process with a structural break. The evidence is particularly strong for seven countries. In most of these countries, the break-trend process is characterized by a shift in the slope of the trend function. The tests fail to reject the unit root hypothesis in the black market real exchange rate of the other eleven countries, thus implying persistent deviations from PPP. Hence the overall evidence on PPP from black market exchange rates is rather mixed. But our results do suggest the need to allow for structural breaks in applications of unit root tests to real exchange rates in countries that have experienced significant shifts in exchange rate/policy regimes.

Appendix: Figures. Black market real exchange rates

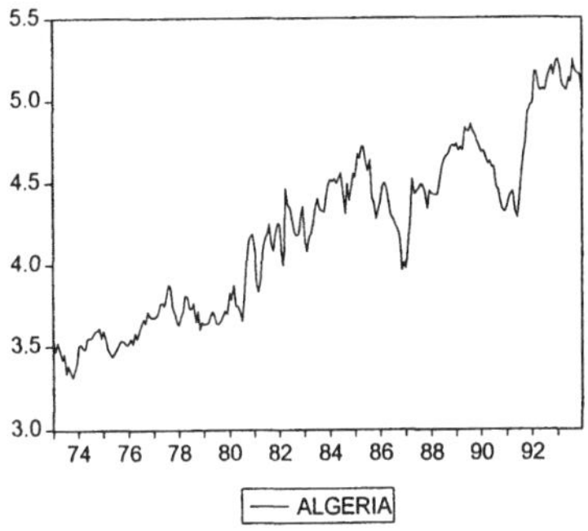

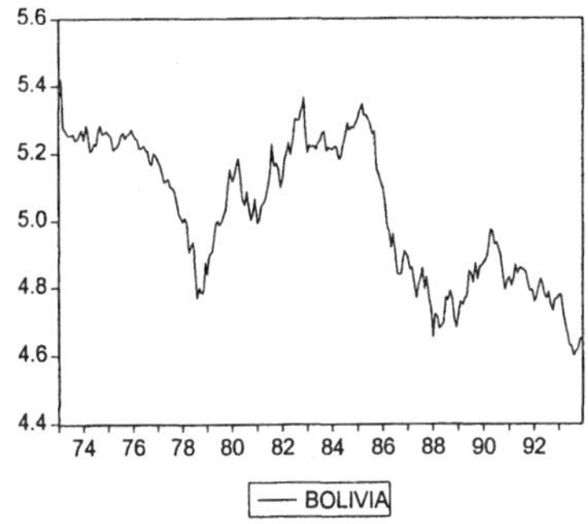

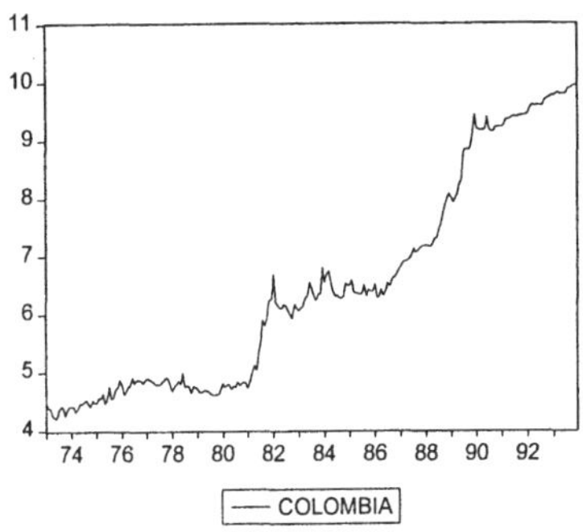

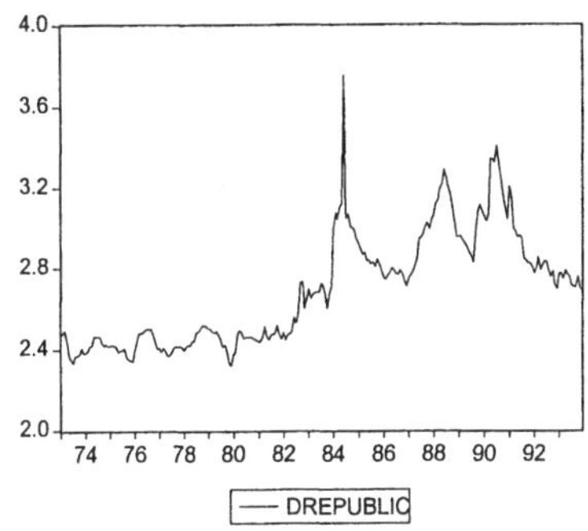

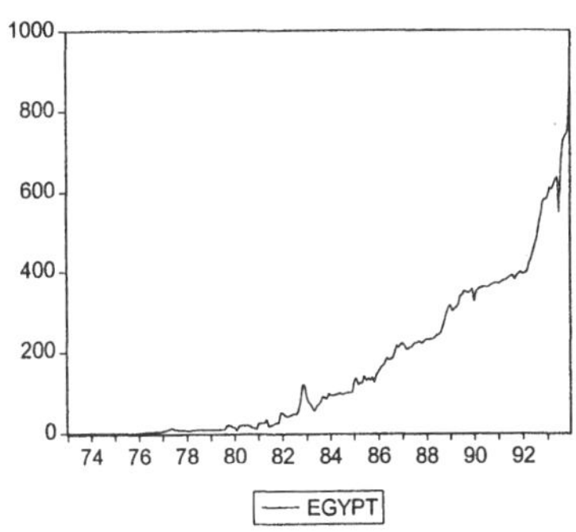

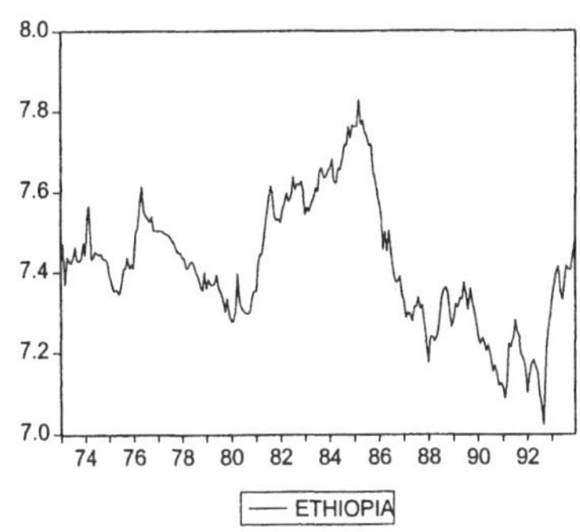

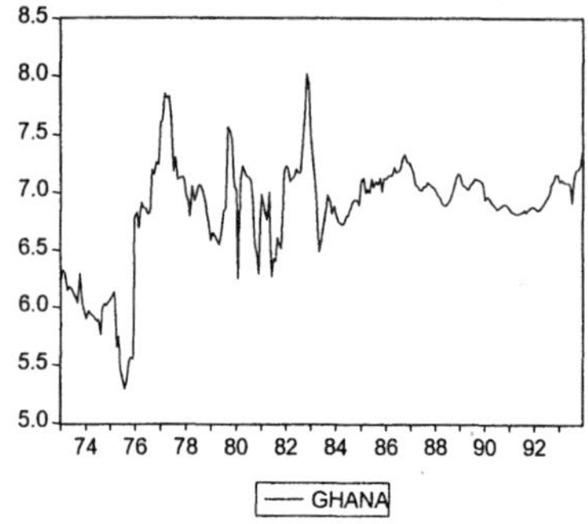
8.5 8.0 7.5 7.0 6.5 6.0 5.5 5.0
74 76 78 80 82 84 86 88 90 92
GHANA

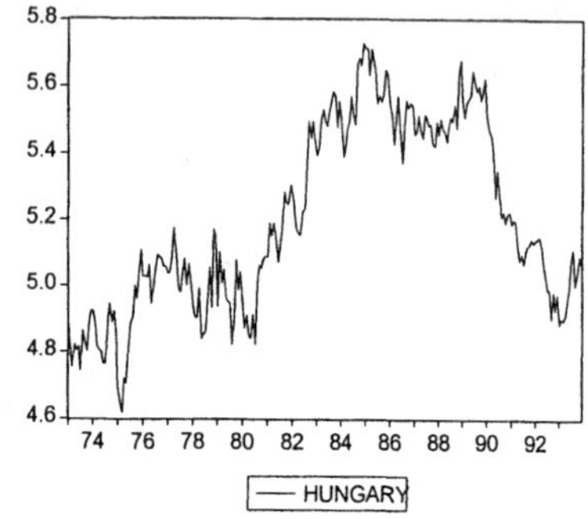
5.8 5.6 5.4 5.2 5.0 4.8 4.6
74 76 78 80 82 84 86 88 90 92
HUNGARY

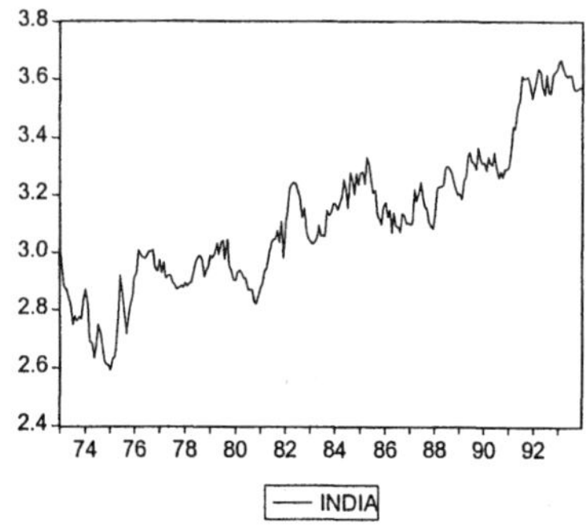
3.8 3.6 3.4 3.2 3.0 2.8 2.6 2.4
74 76 78 80 82 84 86 88 90 92
INDIA

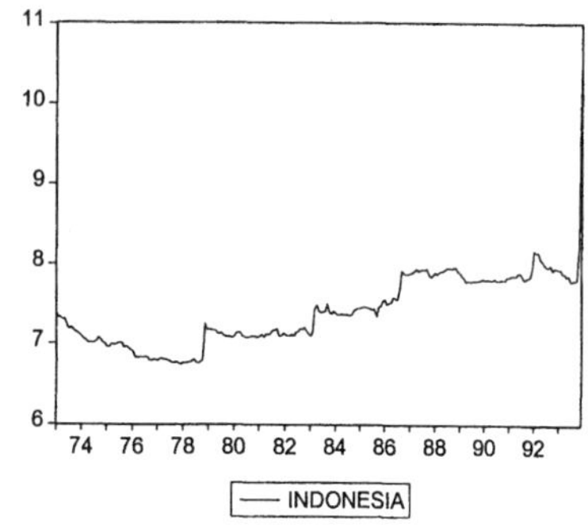
11 10 9 8 7 6
74 76 78 80 82 84 86 88 90 92
INDONESIA

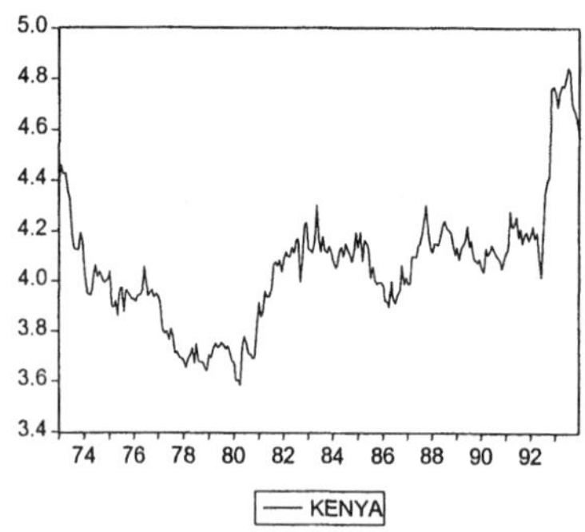
5.0 4.8 4.6 4.4 4.2 4.0 3.8 3.6 3.4
74 76 78 80 82 84 86 88 90 92
KENYA

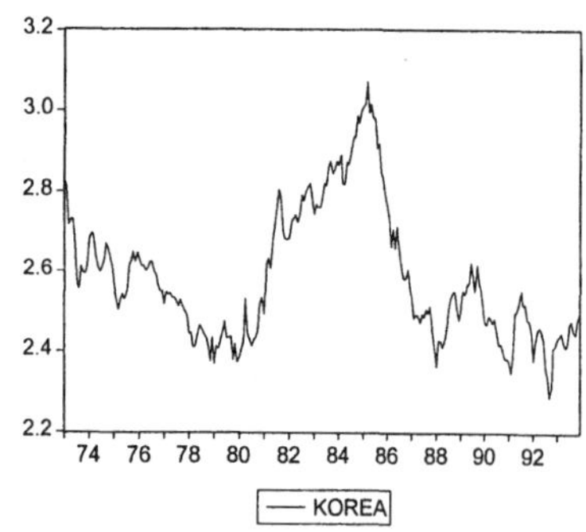
3.2 3.0 2.8 2.6 2.4 2.2
74 76 78 80 82 84 86 88 90 92
KOREA

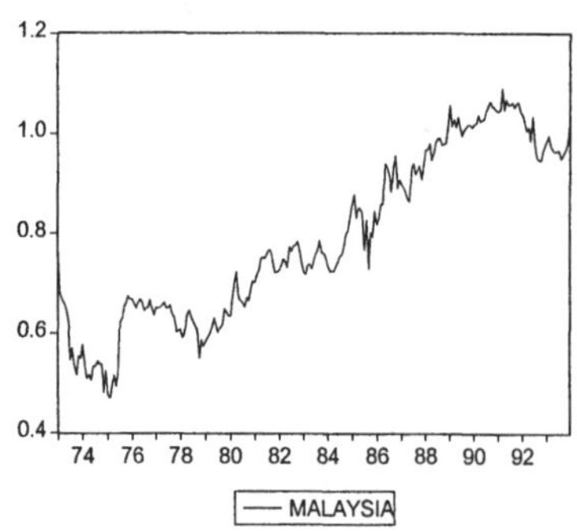
1.2 1.0 0.8 0.6 0.4
74 76 78 80 82 84 86 88 90 92
MALAYSIA

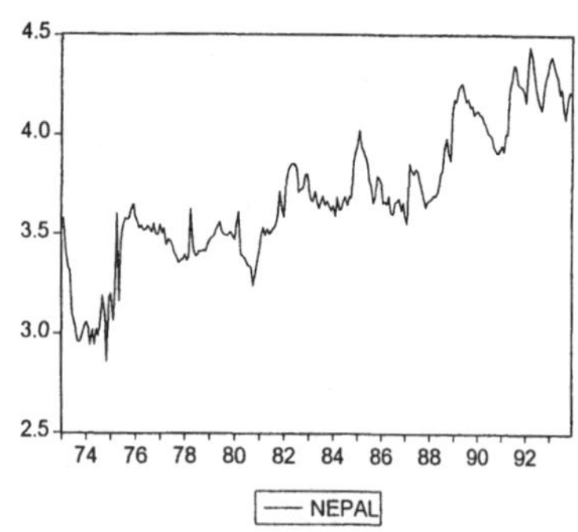
4.5 4.0 3.5 3.0 2.5
74 76 78 80 82 84 86 88 90 92
NEPAL

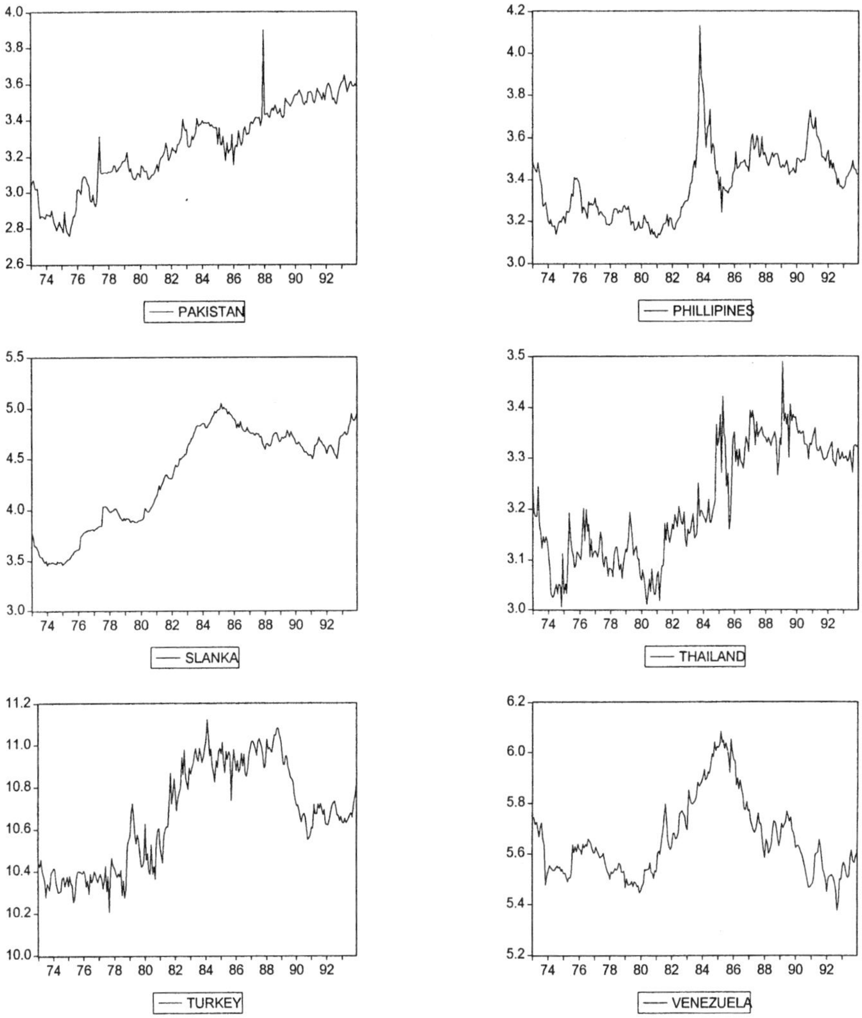

References

Banerjee, A., Lumsdaine, R.L. and Stock, J.H. (1992), "Recursive and sequential tests of the unit root and trend break hypotheses: Theory and international evidence", *Journal of Business and Economic Statistics*, 10, 271-287.

Campbell, J. and Perron, P. (1991), "Pitfalls and opportunities: What macroeconomists should know about unit roots", *NBER Macroeconomics Annual*, 141-201.

Luintel, B.K. (2000), "Real exchange rate behaviour: Evidence from black markets", *Journal of Applied Econometrics*, 15, 161-185.

Papell, D.H. (2002), "The great appreciation, the great depreciation, and the Purchasing Power Parity hypothesis", *Journal of International Economics*, 57, 51-82.

Perron, P. (1989), "The great crash, the oil price shock and the unit root hypothesis", *Econometrica*, 57, 1361-1401.

Phylaktis, K. and Kosimmatis, Y. (1994), "Does the real exchange rate follow a random walk?", *Journal of International Money and Finance*, 13, 476-495.

Sarno, L. and Taylor, M. P. (2002), "Purchasing Power Parity and the real exchange rate", *IMF Staff Papers*, 49, 65-105.

Sen, A. (2000), "On unit root tests when the alternative is a trend-break stationary process", University of Missouri-Rolla, Department of Economics, Working Papers in Economics.

Zivot, E. and Andrews, D.K. (1992), "Further evidence on the great crash, the oil-price shock, and the unit root hypothesis", *Journal of Business and Economic Statistics*, 10, 251-270

Vogelsang, T. J. and Perron, P. (1998), "Additional tests for a unit root allowing for a break in the trend function at an unknown time", *International Economic Review*, 39, 1073-1100.

Part III

Speculative attacks and monetary policy

Chapter 5

The role of foreign speculators during speculative attacks: The case of 1998

By Geir H. BJØNNES[1], Dagfinn RIME[2] and Haakon O.Aa. SOLHEIM[3]

Abstract

We empirically investigate two important issues in the literature on speculative attacks: the role of foreign speculators and the role of contagion between markets. During 1998 both Norway and Sweden experienced a speculative attack, despite having different monetary policy regimes and different economic outlook. The speculative attack can be related to international uncertainty following the debt crisis and the following moratorium issued by Russia in August 1998. We use three sets of explanatory variables: (i) Monetary policy reaction through interest rate setting; (ii) International risk as measured by the 10-year swap spread in the US market; and (iii) Changes in accumulated net holdings of local currency by foreign financial investors. Data on currency holdings are a unique feature of our investigation. We find that during the period of the speculative attack there was a close relationship between holdings of local currency by foreign financial investors, the international risk factor and the exchange rate. This might indicate that contagion between markets works through financial investors. Monetary policy played only a minor role for the outcome of the crisis.

[1] Stockholm Institute for Financial Research (SIFR), and Norwegian School of Management (BI). Email: geir.bjonnes@sifr.org.

[2] Norges Bank (Central Bank of Norway), and SIFR. Email: dagfinn.rime@norges-bank.no.

[3] Norwegian School of Management (BI), and Statistics Norway. Email: haakon.o.solheim@bi.no.

However, the monetary policy regime affected the dynamics of both the exchange rate and the change in holdings of local currency by financial investors.

5.1 Introduction

In this paper we empirically address two largely unresolved issues in the literature on currency crises: the role of foreign speculators and the role of contagion between markets[4].

During 1998 both Norway and Sweden experienced a speculative attack in the foreign exchange market, despite having very different monetary policy regimes and very different economic outlook. The speculative attack can be related to international uncertainty following the debt crisis and the following moratorium issued by Russia in August 1998. To analyze how an external event, like the Russian moratorium, is transmitted to small currency markets like the Swedish krona (SEK/DEM) and the Norwegian krone (NOK/DEM) markets we use two unique data sets. Both Norges Bank (the central bank of Norway) and Sveriges Riksbank (the central bank of Sweden) collect reports of currency trading from market making banks in the local currency markets. From these series we can derive a picture of changes in net currency risk positions of both foreign (financial investors) and local investors.

During 1998 market participants seemed to be very concerned about "international risk". As a measure of "international risk" we use the spread between a 10-year swap in the US market and the 10-year yield on US government bonds. Scandinavian analysts noted that there was a striking similarity between the movement in this risk measures and the local exchange rates during 1998. This risk measure also increased considerably during the Russian crisis. Our results suggest that financial investors, and in particular foreign financial investors, responded to international uncertainty by reducing their currency risk in small currencies. This portfolio shift in turn leads to a depreciation of the local currency. Non-financial participants (*e.g.* exporters/importers or the central bank) take the opposite position.

We find that during the crisis in 1998 changes in portfolios of foreign financial investors became the transmission mechanism of international events into the Scandinavian currency markets. The role of local investors was to take the other side. Furthermore, we find that it is the changes in risk that trigger foreigners in the Swedish market first. Moreover, there is some evidence that changes in financial positions in Sweden affect the NOK/DEM exchange rate, but not the other way around.

It is interesting to view the results in the light of the differences in monetary

[4]Data were kindly provided to us by Norges Bank and Sveriges Riksbank. We are very grateful to Antti Koivisto for helpful discussions and assistance in collecting and correcting the raw data set. Editorial assistance by Janet Aagenæs is appreciated. The views expressed here do not necessarily express those of Norges Bank.

policies. Sveriges Riksbank pursued an inflation targeting regime in 1998, and had no obligation to intervene in support of the exchange rate. Norges Bank was committed to keeping the NOK/DEM rate within a certain (not published) band, although it was made clear that the fluctuation margins would not be strictly enforced. Hence, the NOK followed a managed floating regime. Our results show that the two central banks reacted very differently and that the dynamics of the crisis differed, although movements in both exchange rates were about the same during the crisis. The transmission through foreign financial investors was the same, however, independent of monetary policy regime. In the Swedish market, the change took place more gradually while in the Norwegian market we see a more traditional speculative attack where financial investors speculate against the central bank. Knowing that the central bank will try to defend the currency, financial investors are faced with a one-way bet. Their cost of speculating against the currency is thus primarily related to interest rate differentials.

To our knowledge there are only a very limited number of studies investigating these kinds of questions using data on actual transactions in the currency markets. Most of these studies focus on the role of hedge funds. Wei and Kim (1997) study the importance of hedge funds in currency markets using the Treasury Bulletin reports. They find that trading of hedge funds adds to the volatility of exchange rates, and argue that hedge funds act like "noise traders" in the Korean market. Corsetti, Pesenti, and Roubini (2001) use the same data set, and carry out more informal tests of a number of speculative events. They find support for the role of large foreign investors during speculative attacks. Cai, Cheung, Lee, and Melvin (2001), also using the Treasury Bulletin data, find that trading of hedge funds contributed to volatility during the unwinding of the yen-carry trade in 1998. Our paper is to our knowledge the first contribution that explicitly tries to make an empirical connection between holdings of currency of different groups and contagion between markets.

We follow this introduction with a brief description of our data and the background for this specific event. We then discuss the interaction between the different variables before, during and after the crisis. We end with a conclusion.

5.2 Data and background

In this section we first discuss different aspects of our data. We then give some background on monetary policy regimes and the events in Russia that were influential in triggering the crisis discussed.

5.2.1 Trading data

Norges Bank and Sveriges Riksbank collect data from market making banks on net transactions with different counterparts. From Norges Bank we have access to total net currency transactions over the week divided between foreign and local participants (defined by the address of opposing party), and the central bank. Foreign participants are typically dominated by financial investors. This is especially true in periods of speculative attacks (see Bjønnes, Rime, and Solheim (2002)). In the data set from Sveriges Riksbank we can differentiate between transactions with foreign banks that are not market makers and transactions with non-bank customers. The first group represents the financial investors (see Bjønnes, Rime, and Solheim (2003)). These data are collected on a daily basis. To be able to compare the evidence for Norway and Sweden, we generate weekly series.

The data from Norges Bank distinguish between purchases and sales of spot and forward contracts. Sveriges Riksbank also reports swap and option volumes in addition to purchases and sales of spot and forward contracts. To ensure that we look at actual currency risk, we use series that combine spot and forward positions, as there is some evidence that FX dealers hedge their forward transactions in the spot market[5].

Market microstructure theory suggests that order flow[6] should be correlated with changes in the exchange rate because order flow carries private information. The idea is that investors that believe that the Scandinavian currencies will depreciate will take initiative to buy DEM. However, our data are not similar to order flow data used in most microstructure studies. Note that we distinguish between financial and non-financial customers. All (non-bank) customers will be the initiator of the trades. Rather than focusing on order flow, we examine whether there are systematic differences between different types of customers. To explain this difference, consider the model by Evans and Lyons (2002b). In their model there are three periods. In the first period customers trade with dealers, while in the second period the dealers trade among themselves to share currency risk from the customer trades. In the last period dealers adjust prices to induce customers to absorb their aggregate FX exposure. If FX dealers end the day with zero exposure, net customer trade will be zero measured over the day. However, order flow from the inter-dealer trading will still be correlated with changes in the exchange rate if order flow carries private information. In this study we focus on whether there are systematic differences between the first and the third round customers.

We will test whether financial investors react to changes in international risk by reallocating their portfolios to "safe havens" — in this case large foreign currencies[7]. If this is the case, there should be a positive correlation between the foreign currency holdings and a measure of risk.

Changes in net positions of the market participants that trigger the portfolio shift should be positively correlated with the exchange rate. More specifically,

[5] Note the distinction between *currency risk* and *currency holdings*. A purchase of NOK does not mean that the investor will keep this in cash. Rather the investor is expected to convert his purchase into other assets. However, these assets will be NOK-denominated.

[6] Order flow is signed transactions, given a positive (negative) value for purchases (sales) by the participant taking the initiative to trading.

[7] There is only one foreign currency traded in the Scandinavian interbank currency markets. In 1998 almost all interbank trading was in NOK/DEM and SEK/DEM. Today the euro has taken the role the dmark used to have. Any movements to safe havens like the USD or Swiss franc would then go through the DEM.

if financial investors are triggering portfolio shifts then their positions should be positively correlated with the exchange rate. To induce non-financial (and risk-averse) investors to absorb the portfolio shifts of the financial investors (that is, they take the opposite position), their positions must be negatively correlated with the exchange rate.

5.2.2 The Russian crisis

The trigger event was the Russian crisis. Russia experienced a boom during the mid-1990's, not least due to high oil prices. Oil prices fell from USD 17 at the beginning of 1998 to an average price of USD 12.50 in the period June-October 1998.

In June 1998 Russia began to experience balance of payment problems. Negotiations with the IMF and international creditors were close to breaking down, but on the evening of Sunday, July 12, Russia reached an agreement with the IMF. In the ten days prior to this agreement the central bank of Russia had sold USD 1.6 billion in attempts to stabilize the exchange rate.

The combination of falling oil prices and an unstable government, however, did not allow for the necessary economic reform. During August the crisis reemerged, and on August 17 the Russian president Boris Jeltsin announced a reform package including a possible devaluation of the rouble. The result was a meltdown of international confidence. On August 24 Russia declared a moratorium on all debt payments.

The Russian crisis was only one event in a period of considerable international uncertainty. The East Asian countries were not showing the expected recovery. Uncertainties about world economic growth were increasing. In the months after the Russian moratorium international capital markets fell sharply. The September 17 issue of *The Economist* had the following heading: "Will Brazil be next? - The emerging-market sickness is sizing up a new victim". On October 22 it followed up with a new article, now with the heading "Will China be next?". This clearly indicates the kind of expectations that ruled the markets in this period. Confidence was further shattered as the hedge fund Long Term Capital Management (LTCM) had to be bailed out in October. However, by the end of the year the fear of a total breakdown had receded, and new optimism had restored order in the markets.

5.2.3 International risk factor

Given the discussion above, we need a measure of risk that is exogenous to the markets we look at. One standard measure of such risk is the US swap spread. The spread between the swap rate and the government yield gives an approximation of the cost of taking risk in the US market. When this measure goes up it might be an indication of higher risk also in international markets. This risk measure increased during the summer of 1998, related to the beginning of the turmoil in Russia. It reached its peak at the end of the crisis studied here, and then stabilized at a high level.

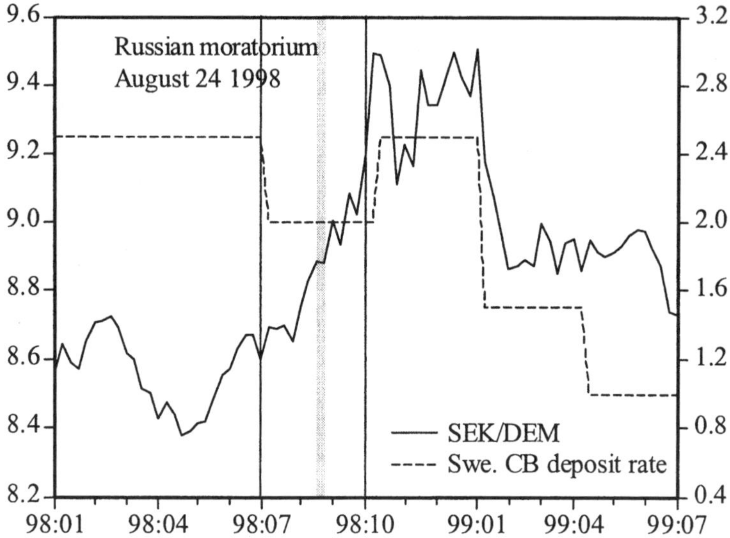

Figure 5.1: Swedish exchange rate and central bank deposit rate.

5.2.4 Monetary policy

Prior to 1992 both Norway and Sweden had fixed exchange rate regimes. However, after speculative attacks during the autumn of 1992 both currencies were allowed to float. The two countries did, however, choose different monetary policy regimes in the aftermath of this flotation.

Sweden chose to adopt a formal inflation target of 2 per cent. In this regime only scant weight was put on stabilizing the exchange rate. Norway, on the other hand, chose a managed float regime. Fluctuation bands were, however, not published[8], and Norges Bank had no obligation to intervene at any specific rate. It was, however, assumed that the central bank would intervene and/or change interest rates if the exchange rate moved "too much".

The reaction functions are partly revealed when we look at figures 5.1 and 5.2. The two thin lines indicate the beginning and the end of the crisis period, July 1, 1998 - October 1, 1998. The grey thick line shows the Russian moratorium on August 24, 1998. We see that while Norges Bank clearly adjusted the deposit rate as the NOK/DEM depreciated in July and August, Sveriges Riksbank did not adjust its deposit rate in response to changes in the exchange rate during this period. This is a clear indication of two very different monetary policy regimes. The large difference in the levels of the deposit rates also

[8]The regulation stated that the rate should be within the band given by the fluctuations of the NOK in the period between the floating and the issue of the regulation (*i.e.* a band defined by the movement of the currency in the period from December 1992 to May 1994).

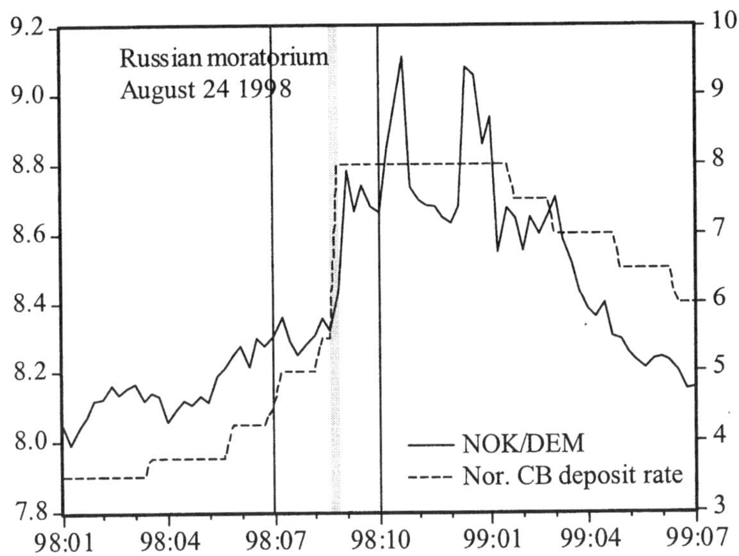

Figure 5.2: Norwegian exchange rate and central bank deposit rate.

indicates that the two countries were in very different economic situations at the start of 1998.

5.3 Results

This section presents our results. We will focus on the period from January 1, 1998 to the end of June, 1999. We split this period into four sub-periods:

1. Pre-crisis: January 1, 1998 - June 30, 1998
2. Crisis: July 1, 1998 - September 30, 1998
3. Turbulence: October 1, 1998 - December 31, 1998
4. Post-crisis: January 1, 1999 - June 30, 1999.

We need to make this distinction to be able to focus on how variables interact during the actual crisis. Pre-crisis and post-crisis periods are set to ensure that these periods are as "normal" as possible. A justification for this split is given in figures 5.1 and 5.2. The first period is one of relative calm in the exchange rate markets. The crisis period is one of abrupt change triggered by the Russian crisis. The period of turbulence is a period of high volatility at a much depreciated level compared to the level in pre-crisis. Although stock markets bottomed out in the beginning of October, uncertainty remained high.

There was also a fear that the crisis could spread to other emerging markets. The post-crisis period is recognized as a period when the exchange rates revert to long-term trends.

As can be seen from the discussion in the previous section, our expectations are as follows:

- There should be a positive correlation between increased international risk and depreciation of the Scandinavian exchange rates.
- There should be a positive relationship between increased international risk and the sale of local currency.
- There should be a positive relationship between the sale of local currency and the depreciation of local currency.

5.3.1 From pre-crisis to crisis to turbulence to post-crisis

Figure 5.3 shows the spread between a 10-year swap in the US market and the 10-year yield on US government bonds, and is a measure of risk. Figure 5.4 shows the log SEK exchange rate, accumulated spot and forward position in foreign currency (non-SEK), against SEK, by foreigners from the beginning of the crisis period. A more positive number reflect larger holdings of foreign currency. Figure 5.5 is similar to figure 5.4, but for the NOK exchange rate and foreign currency against the NOK. Figures 5.3, 5.4 and 5.5 show a strong positive relationship between the swap spread and the two exchange rates, and between positions of financial investors and the exchange rates during the crisis period. We see that the swap spread increases strongly. At the same time, we see that the financial investors sell local currency, while the exchange rate depreciates strongly. Figures 5.3, 5.4 and 5.5 tell more about the difference between developments in NOK/DEM and SEK/DEM in the crisis period. For NOK/DEM we see that changes in the positions of the financial investors and changes in the exchange rate take place during a very short time interval. For SEK/DEM the changes in positions and the exchange rate take place more continuously. As mentioned, this difference is probably related to differences in monetary policy. Furthermore, we see that it is forward flows that are used as a speculative device in the NOK-market, as is usual in a depreciationary crisis, while in the SEK-market the investors sell SEK spot.

Table 5.1 is our first step in investigating this more closely. The table reports regressions of the two exchange rates on changes in positions of financial investors over the four periods defined above[9]. First, we focus on the regression for NOK/DEM. The positions held by financial investors are strongly positively correlated with the exchange rate during the crisis period. In this period foreigners sold NOK, and the currency depreciated. There is thus evidence that financial investors (and in particular foreign financial investors), are triggering

[9] The inclusion of other variables, like short or long interest rate differentials or the oil price for NOK/DEM, does not alter the result. We therefore stick to these simple regressions.

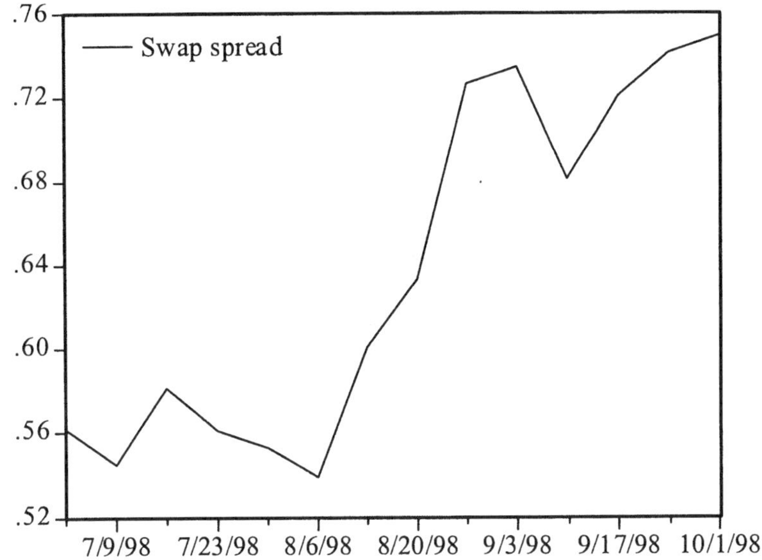

Figure 5.3: Crisis, July 1, 1998 - October 1, 1998. Swap spread.

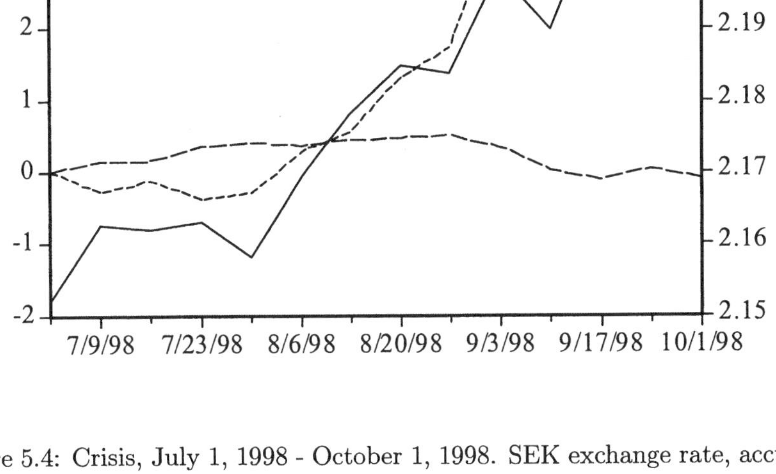

Figure 5.4: Crisis, July 1, 1998 - October 1, 1998. SEK exchange rate, accumulated spot and forward position.

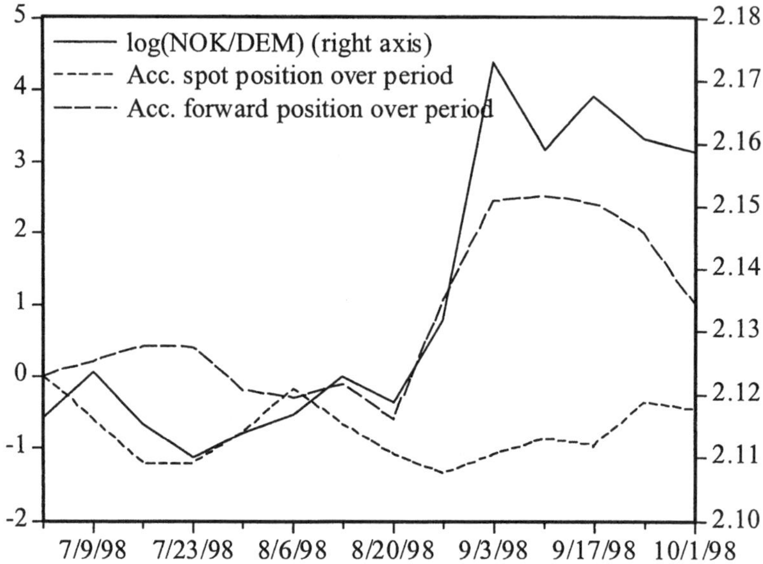

Figure 5.5: Crisis, July 1, 1998 - October 1, 1998. NOK exchange rate, accumulated spot and forward position.

the portfolio shift. In the remaining three periods, we find no positive relationship between positions held by financial investors and the exchange rate. Since Norges Bank was committed to keeping the NOK/DEM within a certain band, one can argue that the net position of a certain group cannot be informative except in periods of speculative attacks. To illustrate this, consider a situation with a fixed exchange rate. To the extent that market participants believe the fixed exchange rate regime will continue to exist, information about net purchases of a particular group will not be informative. If the exchange rate is fixed, there is no currency risk! However, if the credibility of the fixed rate regime is reduced and a speculative attack occurs, there can be a strong correlation between positions of a particular group and the exchange rate.

Next, consider SEK/DEM. Also for SEK we see that the positions of financial investors are positively correlated with the exchange rate in the crisis period. The relationship is significant in both the pre-crisis and crisis periods, but more so in the crisis. This is different from the finding for NOK/DEM. The most plausible explanation originates from differences in monetary policy. With a floating currency, changes in net positions held by a particular group can be informative also in normal periods. If financial investors are responsible for portfolio shifts also in periods other than the crisis period, we can expect to see a positive correlation between their positions and the exchange rate. To induce other types of participants to take the opposite position, prices must be adjusted (if they are risk-averse).

Table 5.1: Regressing exchange rate return on flows for each period

The table shows the regression of changes in log exchange rate on flows in each period. A positive flow reflects a net purchase of foreign currency. Flows are measured as the sum of spot and forward. Flow coefficients measure the impact of a 10 million flow in local currency, approximately USD 1 million, in the SEK and NOK market respectively. ** and * imply significance at the 1% and 5%-level.

	SEK/DEM		NOK/DEM	
	Coeff	*t*-stat	Coeff	*t*-stat
Constant	0.0008	(0.67)	0.0002	(0.19)
Flow (pre)	0.070	*(2.28)	0.013	(1.42)
Flow (crisis)	0.062	**(3.38)	0.123	**(3.25)
Flow (turbulence)	0.092	(0.80)	-0.092	(-1.23)
Flow (post)	0.053	(1.27)	0.019	(0.92)
adjR^2	0.06		0.08	

Pre-crisis: January 1, 1998 - June 30, 1998
Crisis: July 1, 1998 - September 30, 1998
Turbulence: October 1, 1998 - December 31, 1998
Post-crisis: January 1, 1999 - June 30, 1999

Finally, we see that the coefficient for NOK-flows is much larger than the one for SEK-flows. This is an indication that the event was more of a traditional attack in the Norwegian market than in the Swedish market.

Let us turn to the swap spread. We see from table 5.2 that there is a positive correlation between the swap spread and the NOK/DEM exchange rate in all four periods. As expected, the correlation is strongest during the crisis period. As for the SEK/DEM, we see that there is a positive correlation between the swap spread and the exchange rate in two of the periods, the crisis period and the turbulence period. The correlation is strongest during the crisis period.

Table 5.2: Correlation between the local exchange rate and the swap spread

	SEK/DEM	NOK/DEM
Pre-crisis	-0.27	0.47
Crisis	0.91	0.87
Turbulence	0.40	0.55
Post-crisis	-0.11	0.25

Pre-crisis: January 1, 1998 - June 30, 1998
Crisis: July 1, 1998 - September 30, 1998
Turbulence: October 1, 1998 - December 31, 1998
Post-crisis: January 1, 1999 - June 30, 1999

In regressions similar to the ones reported in table 5.1 (not reported here), but using only daily data for SEK/DEM, we find that the swap spread enters with a positive and significant coefficient only in the crisis period.

5.3.2 Transmission by non-financial or financial customers?

An obvious question is the role of financial investors vs. non-financial customers. Evidence can be found in tables 5.3 and 5.4. In table 5.3 we show the correlation between positions of foreign financial investors, local non-financial customers, and the swap spread during the different periods. As can be seen, there is a high positive correlation between the swap spread and positions of foreign financial customers during the crisis period, while the correlation between net currency positions of non-financial customers and the swap spread is negative. During the crisis period we also find a strong negative correlation between financial and non-financial flows in both markets (the correlations are -0.89 and -0.97 in the Norwegian and Swedish market respectively), indicating that non-financial customers provide liquidity in these markets during the crisis.

Table 5.3: Correlation between the swap spread and the net currency positions of locals and foreigners

Table shows the correlation between the swap spread and the net currency positions of local and foreign investors. A more negative number for the holdings reflects a larger holding of local currency.

		Pre-crisis	Crisis	Turbulence	Post-crisis
Norway	Local positions	0.40	-0.43	0.12	0.08
	Foreign positions	-0.56	0.70	-0.17	-0.08
Sweden	Local positions	-0.21	-0.87	-0.08	0.09
	Foreign positions	-0.33	0.93	0.39	0.10

Pre-crisis: January 1, 1998 - June 30, 1998
Crisis: July 1, 1998 - September 30, 1998
Turbulence: October 1, 1998 - December 31, 1998
Post-crisis: January 1, 1999 - June 30, 1999

Table 5.4 adds to the importance of the swap spread, our risk measure, and the role of foreigners. Using the concept of Granger causality on daily data (only available for SEK-flows), we see that (*i*) the swap-spread Granger causes both foreign and local flows, and (*ii*) the foreign flows Granger causes the local flows. Thus, foreigners react to the increased risk by leaving small currencies like the SEK, and hence trigger the crisis.

Table 5.4: Granger causality tests of swap spread and flows. Daily data, July 1, 1998 - October 1, 1998 (Crisis period)

Table shows Granger causality tests for the crisis period for changes in the variables, i.e. change in swap spread and change in foreign currency holdings (spot and forward). Tests include five lags.

	F-Statistic	Probability
Swap spread $\underset{GC}{\not\Rightarrow}$ Foreign flows	4.20	0.00
Foreign flows $\underset{GC}{\not\Rightarrow}$ Swap spread	0.93	0.47
Swap spread $\underset{GC}{\not\Rightarrow}$ Local flows	4.35	0.00
Local flows $\underset{GC}{\not\Rightarrow}$ Swap spread	0.63	0.67
Foreign flows $\underset{GC}{\not\Rightarrow}$ Local flows	2.40	0.05
Local flows $\underset{GC}{\not\Rightarrow}$ Foreign flows	1.28	0.28
Obs.	67	

Where does this leave the Norwegian market? Regression analyses (not reported here) show that the SEK-flows contribute significantly to the NOK/DEM exchange rate both on daily and weekly basis while NOK-flows do not contribute to the SEK/DEM exchange rate. This might indicate that while investors in Sweden were seeking a "safe haven"— a tendency that affected both markets — investors in Norway were more focused on making gains from a speculative attack. So monetary policy did play a role, if not for the final outcome, then at least for the dynamics leading to this outcome.

5.4 Conclusion

In this article we look at the interaction between the exchange rate and three aspects of the currency market: accumulated currency positions, international risk and monetary policy. We find that all three aspects influence the dynamics of the exchange rate. More specifically, we find that net currency positions, international risk and the exchange rate are highly positively correlated during periods of crisis, but are not so systematic in periods before and after such events. This is especially the case for the country with a managed float (Norway). Further, we find that the evident link between international risk and small exchange rates during periods of international uncertainty seems to be driven by change in positions of financial investors, and in particular foreign financial investors. Last, we find evidence that monetary policy does matter for the dynamics of the exchange rate, but the overall pattern in the exchange rate does not seem very affected by the monetary policy regime. This probably reflects the inability (or unwillingness) of governments to pay the cost of short-term stabilization of the exchange rate.

References

Bjønnes, G.H., Rime, D. and Solheim, H.O.A. (2002), " 'Large' vs. 'small' players: A closer look at the dynamics of speculative attacks", mimeo.

Bjønnes, G.H., Rime, D. and Solheim, H.O.A. (2003), "Liquidity provision in the overnight foreign exchange market", mimeo.

Cai, J., Cheung, Y.L., Lee, R.S.K. and Melvin, M. (2001), " 'Once-in-a-generation' yen volatility in 1998: Fundamentals, intervention, and order flow", *Journal of International Money and Finance*, 20(3), 327-347.

Corsetti, G., Pesenti, P. and Roubini, N. (2001), "The role of large players in currency crises", *NBER Working Paper* 8303.

Evans, M.D.D. and Lyons, R.K. (2002a), "Informational integration and FX trading", *Journal of International Money and Finance*, 21(6), 807-831.

Evans, M.D.D. and Lyons, R.K. (2002b), "Order flow and exchange rate dynamics", *Journal of Political Economy*, 110(1), 170-180.

Wei, S.J. and Kim, J. (1997), "The big players in the foreign exchange market: Do they trade on information or noise?", *NBER Working Paper* 6256.

Chapter 6

Exchange rate policy and foreign debt: When is a fear of floating optimal?

By Wolfram BERGER[1]

Abstract

Recent studies have documented that countries with a flexible exchange rate are often reluctant to let their exchange rates float freely. We show, in a model in which a policymaker seeks to stabilize output and inflation around some target values, that the fear of floating is consistent with the policymaker's optimal monetary policy if the relative weight of the output goal remains below a well-defined threshold value. Key features of the model are a credit constraint for the corporate sector, a currency mismatch in firms' balance sheets and a pass-through of exchange rate changes to inflation. The model suggests that even in this environment, flexible exchange rates are an effective tool for absorbing shocks. Monetary policy, however, is less efficient because the costs of the optimal output stabilization rise.

6.1 Introduction

Recent research into exchange rate regimes in developing countries has been dedicated to the question of whether the behavior of exchange rates is consistent with the regimes that governments officially declare they are running. A number of recent studies have documented that developing economies often allow only

[1]University of Hagen, Department of Economics, Universitätsstrasse 41, D-58084 Hagen. Email: wolfram.berger@fernuni-hagen.de. Phone: ++49-2331-987-2416.

This paper was presented at the Applied Econometrics Association Conference on Exchange Rates in Marseille (March 2003). The author is grateful for valuable comments from seminar participants.

Table 6.1: Volatility of nominal exchange rates, foreign exchange reserves and nominal interest rates in selected countries

Country	De jure exch. rate system (period)	NER (±1% band)	FER (±1% band)	NIR (±0.25% band)
		Probability that the monthly percentage change in... falls within a...		
USA	2/1973 - 4/1999	26.8	28.6	59.7
Japan	2/1973 - 4/1999	33.8	44.8	67.9
	Floating			
Bolivia	(9/1985 - 12/1997)	72.8	8.1	16.3
Canada	(6/1970 - 4/1999)	68.2	15.9	36.1
India	(3/1993 - 4/1999)	82.2	21.6	6.4
Mexico	(12/1994 - 4/1999)	34.6	13.2	5.7
Philippines	(1/1998 - 4/1999)	60.7	9.7	22.1
Average floating		51.7	16.2	33.3
	Managed floating			
Bolivia	(1/1998 - 4/1999)	100	6.3	31.1
Brazil	(7/1994 - 12/1998)	83.1	22.2	11.1
Chile	(10/1982 - 4/1999)	45.5	21.3	5
Egypt	(2/1991 - 12/1998)	95.7	37.8	78.9
Indonesia	(11/1978 - 6/1997)	96.4	22.8	30.6
Korea	(3/1980 - 10/1997)	80.1	16.1	31.1
Malaysia	(12/1992 - 9/1998)	59.4	34.3	66.7
Mexico	(1/1989 - 11/1994)	64.3	15.3	8.3
Average managed float		60.1	17.81	36.3

Source: Calvo and Reinhart (2002).
NER: nominal exchange rate, FER: foreign exchange reserves, NIR: nominal interest rates.
Average floating: Average of all countries with a recent or current floating regime, excluding U.S. and Japan.
Average managed float: Average of all countries with a recent or current managed float.

limited nominal exchange rate flexibility despite official announcements that the exchange rate is (by and large) free to float (*e.g.* Calvo and Reinhart (2002))[2]. In Calvo and Reinhart's (2002) words, there seems to be a widespread fear of floating in developing countries[3]. This low flexibility emerges as the result of deliberate policy measures aimed at restricting exchange rate fluctuations and is not owing to the absence of shocks. Some of Calvo and Reinhart's (2002) results are summarized in table 6.1.

[2] In related work, Levy-Yeyati and Sturzenegger (2002) and Reinhart and Rogoff (2002) criticize that the classification of countries' exchange rate systems used by, *e.g.*, the IMF, is based on official announcements and not on an evaluation of what countries really do. In both studies alternative classification schemes based on data on exchange rates and reserves (Levy-Yeyati and Sturzenegger) and on market-determined parallel exchange rates (Reinhart and Rogoff), respectively, are developed.

[3] Based on this result, Calvo and Reinhart conclude that the demise of pegged exchange rates and the trend to corner solutions as argued by, *e.g.*, Fischer (2001) is a myth.

Table 6.2: Relative volatility of nominal exchange rates (averages by country groupings)

Group of countries	Volatility of currency depreciation / volatility of... internat. reserves	interest rate	Internat. reserves in % of M2
G-3	17.55	201.6	6
Other industrialized countries	5.07	38.42	15
Emerging countries	1.76	15.65	37
Other developing countries	0.82	11.06	21
Latin American emerging countries	1.12	9.74	42
East Asian countries	2.63	22.23	39
All countries	4.18	40.57	25

Source: Hausmann, Panizza and Stein (2001).

The first result shown in table 6.1 is not surprising. Exchange rate fluctuations in floating regimes are typically higher than in managed float arrangements. More striking, however, is that the exchange rate flexibility in both regimes is quite low across the countries considered and on average compared to the flexibility of more committed floaters like the U.S. and Japan which serve as a benchmark. The empirical evidence also shows that the volatility, not only of reserves but also of interest rates, is higher relative to countries whose exchange rates are truly floating. Countries seem to stabilize their exchange rate both through direct foreign exchange market interventions and through interest rate policy. Similar conclusions are reached by Hausmann, Panizza and Stein (2001)[4]. Looking both at industrial and developing countries, Hausmann, Panizza and Stein present evidence for a substantial difference in exchange rate policy of countries with floating exchange rate regimes or wide bands. They focus on three different aspects of exchange rate management: The stock of reserves, the relative volatility of exchange rates vis-à-vis reserves, and the relative volatility of exchange rates vis-à-vis interest rates. Basically, the level of reserves and the degree of exchange rate flexibility differ significantly among the countries considered (see table 6.2).

In general, G-3 countries (the U.S., Japan and Germany) allow for a high volatility of the exchange rate relative to that of the level of reserves or of the interest rate while the level of reserves is comparatively low. Developing countries typically limit their exchange rate volatility quite heavily while they hold a high stock of reserves. These results give support to the popular assessment that de-

[4]The results of Calvo and Reinhart (2002) and Hausmann, Panizza and Stein (2001) are confirmed by Levy-Yeyati and Sturzenegger (2002). The latter also point out that the absolute number of countries that fall into the "fear of floating"-category has grown enormously in the 90s. Nevertheless, fear of floating is no new phenomenon. Already in the early to mid 70s and then again in the 80s, a marked discrepancy between official announcements and actual practice in exchange rate policy could be observed albeit only in a comparatively small number of countries.

veloping countries typically seem to be floating "with a lifejacket" (Hausmann, Panizza and Stein (2001))[5].

In this context, a considerable debate about the economic effects of exchange rate changes in developing countries has evolved. The benefits of flexible exchange rates for these economies are seriously called into question. Because of a large share of foreign debt, *i.e.* "dollarized liabilities" (Calvo (2000)), the ability of flexible exchange rates to absorb external shocks might be limited. It is argued that the exchange rate as a policy tool thus loses effectiveness and might even become a potentially harmful instrument. In contrast to the prediction of the standard Mundell-Fleming model, depreciations may be contractionary if serious currency mismatches in the balance sheets of the entrepreneurial sector prevail. If profits are generated in domestic currency while debt is denominated in foreign currency, depreciations may wreak havoc with corporate balance sheets and cause widespread bankruptcies by raising the debt servicing costs in domestic currency[6]. Moreover, the pass-through of exchange rate changes to inflation is generally by far higher in the developing world than in developed countries (Calvo and Reinhart (2000), Goldfajn and Werlang (2000)). Hence, the inflationary consequences of depreciations are more pronounced in developing than in industrial countries. This may constitute another reason why the fear-of-floating-phenomenon is predominantly observed in developing countries.

Starting from the stylized fact that a fear of floating is prevalent in developing economies, we present a model to investigate why and when countries prefer to limit the flexibility of their officially floating exchange rate. Both a currency mismatch in private (*i.e.* firms') balance sheets and a considerable pass-through of exchange rate changes to inflation are key features of the model presented in the following subsection. Against this background we derive a fear of floating as the result of a policymaker's optimal response to real and monetary shocks. We show that a pervasive dollarization of liabilities does not necessarily lead to a fear of floating. We will also touch upon the question of whether flexible exchange rates are still effective in stabilizing output if a large portion of debt is dollarized. This issue has also been the subject of much debate recently (see *e.g.* Céspedes, Chang and Velasco (2001, 2002), Chang and Velasco (2001) and Goldfajn and Olivares (2001)). Our model implies that foreign currency debt does not render the exchange rate an ineffective or even harmful policy tool. But the direction of exchange rate changes in response to shocks may reverse. Adverse shocks might now call for a revaluation of the exchange rate. Our results suggest that dollarized liabilities reduce the efficiency of flexible exchange rates as a shock absorber. Stabilizing the economy against shocks is still possible when firms' liabilities are heavily dollarized but the costs of a given stabilization of output in terms of inflation rise.

---

[5] Hausmann, Panizza and Stein (2001) also present evidence that the ability to float freely is positively correlated with the level of development.

[6] See the theoretical work by, *e.g.*, Aghion, Bacchetta and Banerjee (2000, 2001a, b), Hausmann, Panizza and Stein (2001) and Krugman (1999) and Calvo and Reinhart (2000) for empirical evidence.

The structure of the paper is as follows. In section 6.2 the model is presented. In section 6.3 the optimal monetary policy response to real and monetary shocks is derived and the fear-of-floating phenomenon is explained as the result of the policymaker's optimal stabilization policy. Section 6.4 offers some concluding remarks.

6.2 The model

6.2.1 Output and net worth

In this section a simple stochastic macroeconomic model of monetary policy is developed[7]. We assume a production technology with capital being the only factor. Further, the capital stock fully depreciates within one period. In each period the stock of capital depends on firms' wealth, W_t, which is the sum of their net worth, N_t, and debt, D_t, *i.e.* $W_t = N_t + D_t$. As in, *e.g.*, Hausmann, Panizza and Stein (2001) and Aghion, Bacchetta and Banerjee (2000) a fixed share of firms' total debt is assumed to be denominated in terms of foreign goods[8]. With D_t^f (D_t^d) representing a credit from the international (domestic) capital market total debt in terms of home goods is $D_t = D_t^d + (EX_t/P_t)\, D_t^f$. EX_t is the nominal exchange rate and P_t is the price level so that EX_t/P_t represents the real exchange rate if the foreign price level is normalized to 1. EX_t denotes the price for one unit of the foreign currency expressed in home currency so that an increase in EX_t is a nominal depreciation. Following Bernanke and Gertler (1989), we further assume that due to moral hazard considerations firms face credit constraints[9]. The amount of credit the corporate sector can get depends on its own funds. The higher firms' own funds are, the more they can borrow because the danger of moral hazard behavior decreases. Formally, firms can only borrow an amount proportional to their net worth (which depends on past profits, see below) so that firms' debts can be written as $D_t = \mu N_t$ and firms' total wealth equals $W_t = (1+\mu)\, N_t$[10]. The proportionality factor, μ, is a constant[11]. μ can be interpreted as an indicator for the development of the financial sector (Aghion, Bacchetta and Banerjee (2000)). The higher is μ the

[7] The model is built along the lines of Hausmann, Panizza and Stein (2001) and Aghion, Bacchetta and Banerjee (2000, 2001b).

[8] We do not offer a microeconomic explanation for the currency denomination of debt. The currency denomination of debt is taken as given and not endogenized. Our focus is on the macroeconomic effects of dollarized liabilities, instead. Work by, *e.g.*, Burnside, Eichenbaum and Rebelo (2001), Schneider and Tornell (2003) and Caballero and Krishnamurthy (2003) deals with the micro-causes of the currency denomination of debt.

[9] Credit constraints are more prevalent in developing countries since their financial markets lack the depth of those in industrial countries. Eichengreen and Hausmann (1999) refer to credit market conditions that lead to a currency and maturity mismatch in developing countries as "original sin". The "original sin" leads to a situation in which neither the corporate sector nor the government can use the domestic currency to borrow abroad or to borrow long term, even domestically.

[10] It is assumed that firms borrow the full amount offered by lenders.

[11] Aghion, Bacchetta and Banerjee (2000, 2001b) also consider the case that μ depends negatively on the domestic nominal or real interest rate.

more developed and stable are an economy's credit markets. A small value of μ therefore indicates a pronounced fragility in the financial sector. If μ comes close to zero, the credit market collapses.

Firms retain a fraction of their profits to finance the capital stock needed for production in the following period. Firms' real net worth (in terms of home goods) in the next period $t+1$ is proportionate to the real value of their current period profits,

$$N_{t+1} = \alpha\left(\Omega_t/P_t\right) \tag{6.1}$$

where α denotes the fraction of profits that is saved for investment in the following period. The remaining fraction $(1-\alpha)$ of profits is consumed. Ω_t are the nominal profits of the corporate sector.

Aggregate nominal profits are given by[12]

$$\Omega_t = P_tY_t - \left[EX_t\left(1+i^*\right)D_t^f + \left(1+i_t\right)P_tD_t^d\right] \tag{6.2}$$

where Y_t is output. The interest rate on loans from the global capital market is denoted by i^*, i_t is the interest rate on domestic loans. The foreign nominal interest rate i^* is assumed to be constant in time. Nominal profits are equal to the current nominal income less the production cost which are given by the credit (capital) costs. Firms enter the current period t with the (foreign and domestic) debt burden D_t^f and D_t^d. D_t^f and D_t^d and have to be repaid at the end of period t after monetary policy has been set.

Now, the production technology can be specified. Output is a linear function of the capital stock with which firms enter a period, $Y_t = K_t$, where K_t is linear in firms' wealth and subject to a shock. Taking account of equation (6.1) and the definition of firms' wealth, output is given by

$$Y_{t+1} = \alpha\left(1+\mu\right)\left(\Omega_t/P_t\right)\sigma_t \tag{6.3}$$

We assume that σ_t is stochastic. Specifically, $\sigma_t = \exp\left(\varepsilon_t\right)$ where ε_t represents a white noise shock affecting the productivity of capital. The shock occurs in period t when the capital stock for next period production is accumulated. Accordingly, the output level in $t+1$ is predetermined at the end of period t because firms' net worth and thus their capital stock in period $t+1$ depend on the profits made in period t.

In the following, two periods are considered. Firms enter period 1 with a given stock of domestic and foreign debt. An unanticipated shock occurs and monetary policy (the exchange rate) is adjusted. After the exchange rate has been set, profits for the current period are realized, debt for period 2 is contracted, and output for period 2 is determined through equation (6.3). Moreover, it is assumed that only in period 1 an unanticipated productivity shock occurs.

[12] Remember that the foreign price level is normalized to one.

With the help of equation (6.2), output in period 2 can now be formulated as

$$Y_2 = \sigma_1 \alpha (1+\mu) \left[Y_1 - (EX_1/P_1)(1+i^*) D_1^f + (1+i_1) D_1^d \right] \tag{6.4}$$

Monetary policy in period 1 has an effect on output in period 2 through its impact on firms' profits and therefore on their net worth and their capital stock in period 2. Output in period 1 is exogenous since firms enter period 1 with a predetermined capital stock. Hence, period 1 output cannot be influenced any more by monetary policy actions.

6.2.2 The log-linearized set-up

In this section, a log-linearized set-up is developed. Lowercase letters, except interest rates, denote the logs of the corresponding variables in capitals (the only exception is the exchange rate which is denoted by e_t). We first turn to the evolution of firms' net worth. Aggregate net worth (in terms of home goods) in period 2, n_2, is given by the log-linear approximation in equation (6.5).

$$n_2 = \chi y_1 - (1-\chi) \left(\gamma \left(i^* + e_1 - p_1 + d_1^f \right) + (1-\gamma) \left(i_1 + d_1^d \right) \right) \tag{6.5}$$

χ and γ, where $0 \leq \chi, \gamma \leq 1$, are constants that depend on the steady state[13]. γ denotes the fraction of total debt contracted abroad and reflects the degree of liability dollarization in the economy. Equation (6.5) shows that firms' net worth increases with their income from the previous period and falls with debt servicing costs.

Moreover, uncovered interest parity holds.

$$i_t = i^* + E_{t-1} e_t - e_t \tag{6.6}$$

E_t denotes the rational expectations operator on the basis of all information available in period t. Prices display some stickiness and adjust to changes in the nominal exchange rate.

$$p_t = p_{t-1} + \beta (e_t - e_{t-1}) \tag{6.7}$$

The price level in the current period depends on the price level prevailing in the previous period and it adjusts instantaneously to exchange rate changes[14]. β, where $0 < \beta < 1$, denotes the degree of pass-through of exchange rate changes to inflation. A complete pass-through is precluded. In what follows, the exchange rate in $t-1$ is normalized to zero, $e_{t-1} = 0$.

[13] Actually, log-linearizing equation (6.1) yields: $\tilde{n}_t = \ln \alpha + \ln \Omega_{t-1} - p_{t-1}$. We define $n_t = \tilde{n}_t - \ln \alpha$.

[14] Equation (6.7) is related to the price formulation rule used by Ball (1999). However, in Ball's (1999) formulation, inflation depends on the lagged change in the exchange rate. Moreover, he also considers the impact of lagged output and inflation on the current inflation rate.

Inserting equations (6.6) and (6.7) into the log-linearized version of the production function for period 2 (equation (6.4)) yields[15]:

$$\begin{aligned} y_2 &= \chi y_1 - (1-\chi) \\ &\times \left(i^* + (\gamma(2-\beta)-1)\,e_1 + \gamma\left(d_1^f - p_0\right) + (1-\gamma)\left(E_0 e_1 + d_1^d\right)\right) + \varepsilon_1 \end{aligned} \quad (6.8)$$

A depreciation has two opposing effects on output, an expansionary credit effect and a contractionary balance sheet effect. On the one hand, the costs of domestic credit fall due to the resulting decline in the nominal interest rate and hence, smaller interest payments. Thus firms' net worth rises and output expands. On the other hand, a nominal depreciation, which leads to a real depreciation because a complete pass-through of depreciations on inflation is precluded, *ceteris paribus* increases firms' debt burden in terms of home goods and thus weakens corporate balance sheets. With firms' net worth falling, output contracts. The net-effect of a depreciation on output depends on the relative strength of both effects. Theoretically, the output effect is not unambiguous. If the credit channel dominates the balance sheet channel, depreciations are expansionary, as in typical Mundell-Fleming type models. In the reverse case depreciations are contractionary[16]. More formally, depreciations are expansionary if $(1-\chi)(1-\gamma(2-\beta)) > 0$. The condition for an expansionary effect of depreciations on real activity crucially depends on the degree of liability dollarization, γ, and on the degree of pass-through, β. Depreciations tend to be expansionary if γ is relatively small because then the balance sheet effect of depreciations is rather weak. If γ reaches $1/(2-\beta)$ both effects of exchange rate changes on output exactly cancel out. For a even more pervasive dollarization of liabilities depreciations turn out to have an contractionary output effect. For a given value of γ, depreciations are more likely to have an expansionary effect on output if the degree of pass-through is comparatively high because a given nominal depreciation then results only in a comparatively weak real depreciation, which in turn leads to a rather weak reduction of firms' net worth. While a high pass-through exacerbates the problem of controlling inflation, it mitigates the effects of a depreciation when there is extensive liability dollarization.

The objective function of the policymaker completes our set-up. The policymaker aims at minimizing a standard loss function by optimally setting the exchange rate. The (period) loss function depends on inflation, π_t, and output and is given in equation (6.9).

$$L_t = (\pi_t)^2 + \theta\,(y_{t+1} - \bar{y})^2 \quad (6.9)$$

A monetary policy adjustment in period t determines the current inflation rate and predetermines output in the next period through its effect on firms' current

[15] Actually, log-linearizing the production function yields: $\tilde{y}_{t+1} = \ln\alpha(1+\mu) + n_{t+1} + \varepsilon_t$. We define $y_t = \tilde{y}_t - \ln\alpha(1+\mu)$.

[16] Calvo and Reinhart (2000) present evidence that depreciations in a currency crisis tend to be contractionary and that recessions in the aftermath of a currency crash are by far more severe in emerging markets than in developed countries.

period profits. For simplification, the discount factor is assumed to equal one. As usual, any deviation of inflation and output from their socially efficient target values causes a social loss. The output target, $\bar{y}$, is assumed to equal the steady state value so that there will be no inflation bias. With the help of equations (6.8) and (6.9) the policymaker's objective function can now be formulated as follows:

$$\begin{aligned} L_t &= (\beta e_1)^2 \\ &\quad +\theta \left[\chi y_1 - (1-\chi) \begin{pmatrix} i^* + (\gamma(2-\beta)-1) e_1 + \gamma \left(d_1^f - p_0\right) \\ + (1-\gamma)\left(E_0 e_1 + d_1^d\right) \end{pmatrix} - \bar{y} + \varepsilon_1 \right]^2 \end{aligned} \tag{6.10}$$

The nominal exchange rate, e_1, is the policymaker's policy instrument.

6.3 Dollarized liabilities and the optimal central bank policy

In this section the policymaker's optimal reaction to shocks is derived and discussed. The optimal stabilization policy is characterized by the variance of the exchange rate. To explore whether a fear of floating exists if firms' liabilities are dollarized, the exchange rate variability, measured by the variance of the exchange rate, in the case of dollarized liabilities is compared to the one under the condition that firms' debt is completely domestic.

The reduced form for the exchange rate is

$$\begin{aligned} e_1 &= \frac{\varphi\theta}{\beta^2 + \gamma\theta\varphi(1-\chi)(\beta-1)} \\ &\quad \times \left[(1-\chi)\left(i^* + \gamma\left(d_1^f - p_0\right) + (1-\gamma) d_1^d\right) - \chi y_1 + \bar{y}\right] - \frac{\varphi\theta}{\beta^2 + \theta\varphi^2} + \varepsilon_1 \end{aligned} \tag{6.11}$$

where $\varphi = (1-\chi)(1-\gamma(2-\beta))$[17].

Equation (6.11) shows that the movement of the exchange rate in the presence of foreign debt depends on whether the credit or the balance sheet channel dominates. In comparison to the situation without a shock, an adverse shock ($\varepsilon_1 < 0$) leads to a fall of the exchange rate if the balance sheet channel overcompensates the credit channel ($\varphi < 0$) and causes an increase in the exchange rate if the credit channel overcompensates the balance sheet channel ($\varphi > 0$).

Now, we consider the case where firms borrow from the domestic capital market only. Under this condition, the corporate sector's debt consists totally of domestic loans so that $\gamma = 0$. A deterioration of firms' balance sheets brought

[17] Minimizing equation (6.11) with respect to e_t yields the policymaker's reaction function. After calculating the expected value of e_t and inserting it back into the policymaker's reaction function, the reduced form for the exchange rate given in equation (6.11) can be derived.

about by a (real) depreciation is now precluded. In this case, the policymaker chooses:

$$e_1^d = \frac{(1-\chi)\,\theta}{\beta^2}\left[(1-\chi)\left(i^* + d_1^d\right) - \chi y_1 - \bar{y}\right] - \frac{(1-\chi)\,\theta}{\beta^2 + (1-\chi)^2\,\theta}\varepsilon_1 \qquad (6.12)$$

If balance sheet effect are absent the exchange rate reacts "normally" to shocks. Adverse shocks now unambiguously call for a depreciation. The stabilization policy of the policymaker with and without balance sheets effects can be analyzed by comparing the variance of the exchange rate when the private sector's liabilities are partly dollarized, with the variance of the exchange rate when firms' debt is completely denominated in domestic currency. The exchange rate variance with foreign debt, $\sigma^2_{e_1}$, is $\sigma^2_{e_1} = \left(\frac{\varphi\theta}{\beta^2+\theta\varphi^2}\right)^2 \sigma^2_\varepsilon$ with σ^2_ε denoting the variance of the shock, ε_t, while in the case of domestic debt only, it is $\sigma^2_{e_1^d} = \left(\frac{(1-\chi)\theta}{\beta^2+(1-\chi)^2\theta}\right)^2 \sigma^2_\varepsilon$. The relationship between both variances is not unambiguous but depends on parameter values.

The fluctuations of the exchange rate with and without dollarized liabilities are only equal to each other if θ reaches the threshold value $\bar{\theta}$, where $\bar{\theta} = \frac{\beta^2}{(1-\chi)|\varphi|}$. If, however, θ deviates from $\bar{\theta}$ the debt denomination is a decisive factor for the policymaker's optimal stabilization policy. The policymaker allows greater exchange rate fluctuations when liabilities are not dollarized as compared to the case of dollarized liabilities if the condition $\theta < \bar{\theta}$ is fulfilled. Under these circumstances a fear of floating is the optimal monetary policy in response to shocks. But exchange rate variability may even be higher when firms' liabilities are dollarized. This is the case if θ exceeds the threshold value $\bar{\theta}$.

These results show that the existence of dollarized liabilities and balance sheet effects of exchange rate changes have an impact on the optimal use of the exchange rate as a policy tool. But the model also suggests that the dollarization of corporate balance sheets itself is not sufficient to explain the fear-of-floating phenomenon in developing economies. Why does the policymaker's dislike of exchange rate swings not increase with the amount of foreign currency debt as one could suspect, but vary in dependence on the relative weight attached to her policy goals, instead?

The presence of foreign debt introduces a new channel for the transmission of exchange rate changes on real activity. Through this channel depreciations may wreak havoc with corporate balance sheets and generate widespread bankruptcies. The macroeconomic result is a contraction of real activity. Hence, the balance sheet channel counteracts the usual expansionary output effects of depreciations and renders the impact of exchange rate changes on output ambiguous. The existence of two channels for the transmission of exchange rate adjustments on real activity that work in exactly opposite directions is the basic rationale behind the model's results. To stabilize output at a given level, it is now necessary to offset the counteractive effect of the dominated transmission channel. Hence, exchange rate fluctuations need to be larger than in the case of an unambiguous transmission of exchange rate changes. But greater exchange rate variability to achieve output stabilization results in higher inflation variability since exchange rate changes are passed through to the inflation rate. The gains from stabilizing output are only considered worth these costs if the relative importance of the inflation goal for the policymaker is relatively low, *i.e.* the relative weight of the output goal in the policymaker's objective function exceeds the threshold value $\bar{\theta}$. Hence, a "fear of inflation" (Goldfajn and Olivares (2001)) leads the policymaker to limit exchange rate variability if the inflation goal is sufficiently important relative to the output goal. If, however, the relative weight of the output goal exceeds the threshold value $\bar{\theta}$, exchange rate fluctuations are even larger when firms' debts are denominated in foreign currency. Given the high importance of the output goal, the costs for smoothing output in terms of high inflation volatility are now acceptable for the policymaker. To put it in other word, the "fear of inflation" is not strong enough to prevent the policymaker from stabilizing output through, if necessary, large adjustments of the exchange rate.

The threshold $\bar{\theta}$ that marks the transition from a fear of floating monetary policy to a monetary policy characterized by excessive (relative to the situation without balance sheet effects) exchange rate volatility depends on the model's key parameters γ and β. Neither a variation of γ nor of β has an unambiguous effect on the policymaker's willingness to adopt a fear of floating behavior. Starting from a very low share of foreign debt, an increase in liability dollarization (γ) first makes a fear of floating more likely. While the credit cost channel dominates the transmission of exchange rate changes on output for sufficiently small values of γ, the growing strength of the balance sheet effects induces the policymaker to limit the exchange rate variability out of inflationary considerations. Stabilizing output at a given level requires larger and larger monetary adjustments since a counteracting effect of growing strength has to be offset. This picture reverses if the dollarization of liabilities is so pervasive that the balance sheet channel more than offsets the credit cost channel. Further increases in γ now lead to a decline in the threshold value $\bar{\theta}$ because balance sheet effects increasingly strongly dominate the transmission of exchange rate changes on real activity. Hence, the cost of output stabilization in terms of inflation (or deflation, respectively) decline and the policymaker is willing to tolerate larger exchange rate swings.

The model's second key parameter, the degree of pass-through β, not only has a direct impact on the inflation rate but it also determines the comovement between nominal and real exchange rates. If the latter is comparatively low due to a considerable pass-through, nominal exchange rate changes only have a minor effect on firms' real net worth. A weak pass-through (and thus a high commovement between the nominal and the real exchange rates), however, enhances the impact of (nominal) exchange rate fluctuations on firms' balance sheets so that the required adjustments of the exchange rate are relatively small. Hence, the smaller is β the less likely is the exchange rate volatility restricted due to a fear of inflation. If β gets larger, the adjustment of the nominal exchange rate needs to be larger in response to shocks. The impact of the nominal exchange rate on corporate balance sheets declines since the correlation between nominal and real exchange rate movements falls. Thus, the policymaker is more and more inclined to adopt a fear of floating behavior out of concern about inflation variability. If β is sufficiently large, the price effects of exchange rate changes are so strong that the credit cost channel becomes the dominating transmission channel. Now, further increases in β lead to a fall in the extent of exchange rate changes needed to absorb a given shock. Thus, the costs of stabilizing output in terms of inflation (or deflation, respectively) tend to fall. But at the same time, a larger pass-through compounds the problem of volatile inflation brought about by exchange rate fluctuations. On the whole, for sufficiently large β the likelihood for a fear of floating is quite small but the policymaker's willingness to restrict exchange rate movements rises steadily.

Our results lead us to the conclusion that a pervasive dollarization of liabilities does not necessarily render a flexible exchange rate an ineffective shock absorber[18]. The exchange rate is still effective in stabilizing the real economy against shocks according to the policymaker's preferences. The optimal stabilization policy in response to a given shock may now be achieved through a depreciation or a revaluation of the exchange rate depending on the relative strength of the transmission channels. But compared to the case without dollarized liabilities, larger exchange rate changes are required to restore a given level of real activity after a shock has hit since the counteractive effect of the dominated transmission channel has to be offset. Hence, the costs of stabilizing output in terms of inflation rise. This key result of the model can be interpreted as showing that the *efficiency* of the exchange rate as a policy tool is reduced through the dollarization of liabilities.

6.4 Conclusion

Recent studies have documented that countries with a flexible exchange rate are often reluctant to let their exchange rates float freely. In particular, among the developing countries it seems to be common practice to limit the fluctuation of an exchange rate officially labeled as 'flexible'. An often-cited culprit for this stylized fact is a currency mismatch in corporate balance sheets. Firms' profits are denominated in home currency while their debt is often largely denominated in a foreign currency. Depreciations may now be contractionary because they increase firms' debt burdens in the home currency and may thus lead to widespread bankruptcies. Hence, the effect of exchange rate changes on output well-known from the standard Mundell-Fleming approach may reverse.

We show, in a model in which a policymaker seeks to stabilize output and

[18] This conclusion is also reached in other work. Céspedes, Chang and Velasco (2002) show in a dynamic optimizing model that the insulating role of flexible exchange rates against real foreign shocks, which is a central result of Mundell-Fleming type models, is still given in a small open economy where financial frictions lead to a credit constraint for the corporate sector and liabilities are dollarized (see also Chang and Velasco (2001) for a similar result). In a related model, Céspedes, Chang and Velasco (2001) explore the optimal monetary and exchange rate policy when balance sheets matter. They reach the conclusion that fixing the nominal exchange rate is inferior to an inflation targeting policy with flexible exchange rates.

inflation around some target values, that a fear of floating can be derived as the by-product of the optimal stabilization policy of a policymaker who cares about output and inflation. The policymaker may prefer to restrict exchange rate fluctuations if there are considerable currency mismatches in corporate balance sheets because achieving the optimal output stabilization requires larger exchange rate movements than in the case without these mismatches. The reason for this result is that, in the presence of foreign debt, exchange rate fluctuations are transmitted through two channels that work in opposite directions. To stabilize the economy the impact of the dominated transmission channel has to be offset first. This puts at risk a good inflation performance so that the exchange rate is not allowed to float as freely as it would in an environment without balance sheet effects. In this case, a fear of inflation determines the policymaker's monetary policy and gives rise to a fear of floating.

Dollarized liabilities do not, however, limit exchange rate flexibility in every case. Our model shows that it may even be optimal for the policymaker to allow greater exchange rate fluctuations if corporate liabilities are denominated in foreign currency. If the output goal becomes increasingly important relative to the inflation goal, the fear of inflation analyzed above does not restrict exchange rate fluctuations. The costs brought about by large deviations of the inflation rate from its target are overcompensated by reducing the output costs through, if necessary, large adjustments of the exchange rate.

The model suggests that flexible exchange rates and a countercyclical monetary policy can buffer shocks even in an economy with a credit constrained corporate sector and widespread dollarization of liabilities. Depending on the dominating channel for the transmission of exchange rate changes on real activity, offsetting an adverse shock may require a depreciation or a revaluation of the exchange rate. But the presence of two counteractive transmission channels leads to an increase in the costs of output stabilization since achieving a given output level in response to shocks now requires larger exchange rate movements. Hence, the model suggests that the efficiency of monetary policy in stabilizing the economy falls with an increase in the dollarization of liabilities.

References

Aghion, P., Bacchetta, P. and Banerjee, A. (2000), "A simple model of monetary policy and currency crises", *European Economic Review,* 44, 728-738.

Aghion, P., Bacchetta, P. and Banerjee, A. (2001a), "A corporate balance-sheet approach to currency crises", *Working Paper* 01.05, Studienzentrum Gerzensee.

Aghion, P., Bacchetta, P. and Banerjee, A. (2001b), "Currency crises and monetary policy in an economy with credit constraints", *European Economic Review,* 45, 1121-1150.

Ball, L. (1999), "Policy rules for open economies", *in* J. Taylor (ed.), *Monetary policy rules*, University of Chigaco Press, Chicago, 127-144.

Bernanke, B. and Gertler, M. (1989), "Agency costs, net worth, and business fluctuations", *American Economic Review,* 79, 14-31.

Burnside, C., Eichenbaum, M. and Rebelo, S. (2001), "Hedging and financial fragility in fixed exchange rate regimes", *European Economic Review,* 45, 1151-1193.

Caballero, R. and Krishnamurthy, A. (2003), "Excessive Dollar debt: Financial development and underinsurance", *Journal of Finance,* 58, 867-893.

Calvo, G. (2000), "Capital flows and the exchange rates with special references to the Dollarization in Latin America", http://www.bsos.umd.edu/econ/ciecrp9.pdf (accessed April 2003).

Calvo, G. and Reinhart, C. (2000), "Fixing for your life", *NBER Working Paper* 8006, Cambridge, MA.

Calvo, G. and Reinhart, C. (2002), "Fear of floating", *Quarterly Journal of Economics,* 117, 379-408.

Céspedes, L., Chang, R. and Velasco, A. (2001), "Dollarization of liabilities, net worth effects, and optimal monetary policy", http://ksghome.harvard.edu/~.AVelasco.Academic.Ksg/islafinal.pdf (accessed April 2003).

Céspedes, L., Chang, R. and Velasco, A. (2002), "Balance sheets and exchange rate policy", http://ksghome.harvard.edu/~.AVelasco.Academic.Ksg/Files/Research/versionjune2002.pdf (accessed April 2003).

Chang, R. and Velasco, A. (2001), "Monetary policy in a Dollarized economy where balance sheets matter", *Journal of Development Economics,* 66, 445-464.

Eichengreen, B. and Hausmann, R. (1999), "Exchange rates and financial fragility", *NBER Working Paper* 7418, Cambridge, MA.

Fischer, S. (2001), "Exchange rate regimes: Is the bipolar view correct?", *Journal of Economic Perspectives,* 15, 3-24.

Goldfajn, I. and Olivares, G. (2001), "Can flexible exchange rates still "work" in financially open economies?", UNCTAD G-24 Discussion Paper No. 8, New York.

Goldfajn, I. and Werlang, S. (2000), "The pass-through from depreciation to inflation: A panel study", *Working Paper* No. 423, PUC Rio de Janeiro, Rio de Janeiro.

Hausmann, R., Panizza, U. and Stein, E. (2001), "Why do countries float the way they float?", *Journal of Development Economics,* 66, 387-414.

Krugman, P. (1999), "Balance sheets, the transfer problem, and financial crises", in: P. Isard, A. Razin and A. Rose (eds.), *International finance and financial crises, Essays in honour of Robert. P. Flood, Jr.*, Kluwer, Dordrecht, 31-44.

Levy-Yeyati, E. and , F. Sturzenegger (2002), "Classifying exchange rate regimes: Deeds vs. words", http://www.utdt.edu/~fsturzen/DW2002.pdf (accessed April 2003).

Reinhart, C. and Rogoff, K. (2002), "The modern history of exchange rate arrangements: A reinterpretation", *NBER Working Paper* 8963, Cambridge, MA.

Schneider, M. and Tornell, A. (2003), "Balance sheet effects, bailout guarantees and financial crises", http://www.econ.ucla.edu/schneider/sc57.pdf.

Chapter 7

Learning and coordination against a fixed exchange rate

By Christophe CHAMLEY[1]

Abstract

A canonical model of coordination is presented in which a large number of agents gather over time information about the mass of potential participants by the observation of others' actions. An agent who delays receives more information but incurs the risk of missing sharing in the spoils of a victorious coordination. Applications can be found in various settings including speculative attacks against a regime of completely fixed exchange rate. The model complements that of Chamley (2003b) in which the exchange rate is allowed to fluctuate within a band.

7.1 Introduction

A regime of fixed exchange rates may be toppled by speculators if they coordinate on selling the currency and their sales exceed the reserves of the central bank. A simple static game of one period has been proposed by Obstfeld (1996): under perfect information on the strategies and the payoffs of individuals, there are two equilibria in pure strategies, as in any standard stag hunt game: either speculators cooperate and gain, or they do not in which case there is neither gain nor loss. A critical factor for the success of a speculative attack is the distribution of characteristics of the speculators about which the central bank may not have superior information. The one-period approach leaves little room for the role of the central bank during a speculative attack. Furthermore, speculative attacks do no take place in one period. Agents observe intently what other

[1]Boston University and EHESS-DELTA. Boston University, 270 Bay State Road, Boston, MA, 02215.

do during an attack and use that information to evaluate whether an attack is likely to succeed and worth pursuing.

These issues are addressed here with a model where agents act strategically in a multi-period framework with a regime of fixed exchange rate. The model has canonical properties and can be applied to a variety of contexts. Any speculator in a currency attack wonders how many other agents believe that the fundamentals are such that an attack will succeed, how others think about others' beliefs, and so on. This feature is modeled here by assuming that the mass of speculators is an unobservable parameter which is fixed randomly at the beginning of time. For simplicity, there are two states of nature. In one of the two states, the "high" state, the mass of speculators is sufficiently large to induce a devaluation if all speculators buy the foreign currency while that mass is "sub-critical" in the other "low" state.

The emphasis is not on the resolution of multiple equilibria as in the global game approach (Carlsson and van Damme (1993a, 1993b))[2], but on the opportunities offered by the dynamic setting for speculative attacks when agents have imperfect information on their total mass. In all the cases considered here, there will be an equilibrium with no speculative attack. Under some conditions, there will also be an equilibrium (or many equilibria) with a speculative attack. The role of policy, if any, will be to abolish any equilibrium of the second type. The strategic aspect is critical: the interesting cases occur when the parameters of the model are such that if there is one-period (and thus no opportunity to learn from others), there is a unique equilibrium with no attack[3].

In a previous paper (Chamley (2003b)), I show how agents with imperfect information may interact strategically when the exchange rate is allowed to fluctuate in a narrow band, as in European Monetary System before 1999. In each period speculators face a trade-off between selling the currency now at a relatively favorable price and waiting for more information but at the risk of a lower price of the currency and a smaller probability of trading before the devaluation takes place. Within the band, the exchange rate is determined as in a standard financial market with three types of agents: speculators are more optimistic about a devaluation than other agents and place market orders. The other two types of agents are noise traders and market-makers. Speculators know the structure of the market, hence the exchange rate (as long as it is within the band), is informationally equivalent to the order flow. Because the strategy of speculators is known, the order flow is a noisy signal about the mass of speculators participating in the attack. The signal to noise ratio increases with the amount of trading by each speculator.

[2] Morris and Shin (1999) apply the global game method to speculative attacks in a multi-period framework, but there is no endogenous learning nor option to delay in their model.

[3] Gale (1995) has shown how a finite number of agents may achieve coordination in multiple periods when each agent can precipitate a subgame with fewer agents in which coordination is dominant. Rodionova and Surti (2000) analyze the case of speculative attacks with a finite number of players and Corsetti *et al.* (1999) assume one large agent who can induce small and competitive agents to act.

In this paper, I assume that the exchange rate is perfectly fixed with no band of fluctuation and that speculators have direct information about the trade flow. Agents face a fixed transaction cost and are heterogeneous: individual costs are distributed on an interval. Agents play against a fixed level of reserves by the central bank. They do not know whether their total liquidities are greater or smaller than the reserves of the central bank. In the latter case, any speculative attack is doomed.

Payoff externalities have a particular property here. When speculators face a coordination problem (*e.g.* attack a currency), there is some incentive to delay in order to get more information on others[4]. But there is also a premium for the agents who invest earlier in the coordination: if the agent delays too much, he comes too late for the gain of the attack and does not get anything. In an equilibrium with a speculative attack, agents with a lower cost go early, as in the model Chamley (2003a) which has no payoff externalities. Agents with a relatively higher cost delay in order to get more information about the mass of agents who can buy the foreign currency and face the risk of "getting to the window" too late: they could miss the devaluation this period, or if they get to place an order in the future, they may be rationed by the fixed amount of reserves of the central bank. The model incorporates a trade-off between "going early" with little information, and delaying for more information with the risk of coming too late. In all periods, the subgame has an equilibrium in which there is no speculation for all remaining periods. Under some conditions on the beliefs of the agents, there are other equilibria with speculative attacks. A higher number of periods extends the set of beliefs for which a speculative attack is an equilibrium. The equilibrium strategy which is analyzed here converges to a stationary solution when the number of periods tends to infinity. A numerical example with Gaussian noise illustrates some properties of the model.

Policies are analyzed in Section 7.4. In the context of the model, the central bank can prevent any speculative attack by raising the discount rate or by adding a component to the amount of trade. If this trading by the central bank is observable, it is ineffective. But if the central bank's trade is not observable it adds to the noise and reduces the information content of the trade flow. The reduction of information may be sufficient to prevent a speculative attack. The remarkable property is that the reserves which are required for such a policy may be smaller than the reserves which would be required for a successful defense of the exchange rate if speculators had perfect information.

7.2 The model

There are $T + 1$ periods and a continuum[5] of speculators, called agents, of mass $\theta \in \{\theta_0, \theta_1\}$. The value of θ defines the state which is selected by nature

[4] The incentive for delay generates strategic substitutability in Chamley and Gale (1994), and strategic complementarity when agents are heterogeneous in Chamley (2003a).

[5] Dynamic speculative attacks with pivotal agents are analyzed in Rodionova and Surti (1999).

before the first period. Each agent holds one option to take one action in any period of his choice. This action can be an investment or the purchase of a fixed amount of foreign currency which is normalized to one. When an agent takes his action and "invests" he incurs a fixed cost c which is idiosyncratic. The costs of individual agents are uniformly distributed on an interval $(0, \bar{c})$ where $\bar{c}$ is a fixed parameter.

The structure of the game is general, but we will present it in the context of a regime of fixed exchange rate, normalized to one, between the domestic and the foreign currency. At the beginning of time, each agent holds one unit of domestic currency and this amount embodies a liquidity constraint. There is no loss of generality if we assume that before period $T+1$, each agent buys the foreign currency, called *asset*, at most once in which case he sells his entire amount of domestic currency, called *currency*. For simplicity, it is assumed that the cost of transaction is incurred only once when the agent buys the asset.

In period $T+1$, all agents undo their position: by assumption, all agents must hold only the currency at the end of period $T+1$; if an agent holds the asset at the beginning of period $T+1$, he sells it in period $T+1$. The game is effectively played during T periods.

There is a game in period t if no devaluation (a process described below) has taken place before period t. Let λ_{t-1} be the fraction of agents who hold the asset at the beginning of period t. The mass of agents who hold the asset at the beginning of period t is therefore $\lambda_{t-1}\theta$. By assumption, no agent holds the asset at the beginning of the first period: $\lambda_0 = 0$.

We will consider only symmetric monotone strategies[6]: each currency holder invests if and only if his cost is smaller than $\zeta_t \bar{c}$ with $\zeta_t \in [0, 1]$. The value of ζ_t will define the strategy of currency holders. The strategy of the asset holders will be simpler as we will see later: if some agents buy in an equilibrium, no asset holder sells, and if some asset holders sell, they all sell.

Suppose that the attack goes on in period t: more agents buy the asset and $\zeta_t > 0$. Given the quantity of orders $\zeta_t\theta$, the demand for the asset by the agents in period t (a stock) is $(\lambda_{t-1} + \zeta_t)\theta$. The total demand for the asset in period t, y_t, is the sum of the agents' endogenous investments and of an exogenous noise, η_t. The noise traders in period t undo their position at the beginning of the next period. The distribution of η_t is stationary. The terms η_t are serially independent and the mean of η_t is equal to zero.

There is a fixed amount $\bar{X}$ of reserves in the central bank. This value is perfectly known by all agents[7].

If $y_t \leq \bar{X}$, the total investment is smaller than the reserves of the central bank. There is no devaluation in period t. All investment orders are satisfied. The fraction of agents who hold the asset at the beginning of the next period is $\lambda_t = \lambda_{t-1} + \zeta_t$.

If $y_t > \bar{X}$, the demand for the asset exceeds the reserves of the central bank. A devaluation takes place. All new orders in period t cannot by executed and

[6] The monotone property is not restrictive, as may be shown by the reader using arguments similar to those in Chamley (2003a).

[7] The analysis of model with an imperfectly known level of reserves remains to be done.

some rationing must take place. To simplify the process, it is assumed that: first, noise traders of period $t-1$ undo their positions; second, noise traders of period t place their investment orders. All these orders are executed (even if the total amount exceeds the critical mass)[8]. The amount of the asset which is available for new orders without devaluation is therefore $Max(\bar{X}-\lambda_{t-1}\theta-\eta_t, 0)$. (We will see later that if new orders come in, no asset holder sells).

Suppose that the mass of new orders in period t is strictly positive: $\zeta_t > 0$. (The case $\zeta_t = 0$ will be described below). By assumption, all the agents' new orders are executed with the same probability which must be

$$\pi = \frac{Max(\bar{X} - \lambda_{t-1}\theta - \eta_t, 0)}{\zeta_t \theta}.$$

By construction, $0 \leq \pi \leq 1$. If a devaluation takes place, the price of the foreign asset is set at $1 + A$, where $A > \bar{c}$ is a fixed parameter[9]. In this case, the game ends. All agents holding the asset realize their capital gain and purchase the domestic currency.

Agents have a discount factor $\delta \leq 1$ between periods. Note that the discount factor may be equal to one. In standard models with delays (*e.g.*, Chamley and Gale (1994)), the discount factor must be strictly smaller than one in order to generate a cost of delay and induce some agents to go early. No such assumption is required here: the cost of delay is generated by the risk of missing the devaluation. However, a discount factor is introduced for the sake of generality and for later policy analysis.

Information

The true state θ is not observable. At the beginning of the first period, all agents have a subjective probability μ_0 of the high state θ_1. At the end of each period t, if a devaluation takes place, the game ends. If no devaluation takes place, agents observe the total quantity of investment y_t which is given by

$$y_t = (\lambda_{t-1} + \zeta_t)\theta + \eta_t. \tag{7.1}$$

We will show that the subgame which begins in period t depends only on the period t and on $(\lambda_{t-1}, \mu_{t-1})$ which measures at the beginning of the period the fraction of agents holding the asset and the level of belief (the probability of state θ_1).

[8] This assumption is made to simplify the process. The probability of a demand greater than the critical mass at that stage is very small.

[9] One could consider the case where the amount of the devaluation is determined by the intensity of the attack, or by an equilibrium mechanism. Such an effect would enhance the strategic complementarity but would not alter the properties of the model.

7.2.1 Equilibria

The evolution of beliefs

Since the strategies are common knowledge, the observation of the demand y_t which is defined in (7.1) is equivalent to the observation of the variable

$$z_t = \theta + \frac{\eta_t}{\lambda_{t-1} + \zeta_t}. \tag{7.2}$$

In this expression, the variance of the noise term is reduced when the demand by speculators is higher: the information conveyed by the trades (conditional on no devaluation) increases with the fraction of agents who place orders. Let $f(\eta)$ be the density of η (which is independent t). If there is no devaluation in period t, the belief on θ_t in the next period is determined by the Bayesian updating formula

$$\frac{\mu_t(y_t; \lambda_{t-1} + \zeta_t, \mu_{t-1})}{1 - \mu_t(y_t; \lambda_{t-1} + \zeta_t, \mu_{t-1})} = \frac{\mu_{t-1} f(y_t - (\lambda_{t-1} + \zeta_t)\theta_1)}{(1 - \mu_{t-1}) f(y_t - (\lambda_{t-1} + \zeta_t)\theta_0)}. \tag{7.3}$$

Payoff of an order

From the description of the trades, a devaluation takes place in period t if $y_t > \bar{X}$, in which case the *ex post* gross payoff of an order is $A\frac{Max(\bar{X} - \lambda_{t-1}\theta - \eta, 0)}{\zeta_t \theta}$.

We define all payoff as differences with the payoff of holding the currency with no trade. If no devaluation takes place in period t, the *ex post* gross payoff is the value of holding the asset in the continuation of the game which is $\delta\, V_{t+1}(\lambda_{t-1} + \zeta_t, \mu_t)$. The function $V_{t+1}(\lambda_{t-1} + \zeta_t, \mu_t)$ is the value of holding the asset at the beginning of period $t+1$. It depends on the fraction of agents holding the asset at the beginning of that period, $\lambda_t = \lambda_{t-1} + \zeta_t$ and on the belief μ_t which has been expressed in (7.3).

The payoff of an order is the expected value of all *ex post* payoffs for all possible values of θ and η. We denote by $F(\theta, \eta; \mu_{t-1})$ the *c.d.f.*[10] of (θ, η) at the beginning of period t. The strategy ζ_t defines the cost of the marginal agent, $(\lambda_{t-1} + \zeta_t)\bar{c}$. We assume that the transaction cost is incurred whether the order is executed or not. (The case of a cost contingent on the execution of the order would be similar). The net payoff of this agent is

$$\begin{aligned} u_t\left(\zeta_t; \lambda_{t-1}, \mu_{t-1}\right) &= -(\lambda_{t-1} + \zeta_t)\bar{c} \\ &\quad + \int_{y_t > \bar{X}} A\frac{Max(\bar{X} - \lambda_{t-1}\theta - \eta, 0)}{\zeta_t \theta} dF\left(\theta, \eta; \mu_{t-1}\right) \\ &\quad + \delta \int_{y_t < \bar{X}} V_{t+1}(\lambda_{t-1} + \zeta_t, \mu_t) dF\left(\theta, \eta; \mu_{t-1}\right). \end{aligned} \tag{7.4}$$

[10] θ and η are independent; the distribution of θ depends on the belief μ_{t-1}, and the distribution of η_t is independent of the period.

The method of backward induction which will be used here can characterize all equilibria of all subgames but such a complete characterization is beyond the scope of this paper. We will assume that if in period t there is no equilibrium strategy with $\zeta_t > 0$ (no new order comes in), then the speculative attack stops completely[11]. One can show that such an outcome is an equilibrium in this model (Chamley (2003b)). In the subgame which begins in period t, there may be multiple equilibrium values for $\zeta_t > 0$. (An example will be given below). When there are such multiple equilibrium strategies, we assume that agents coordinate on the highest equilibrium value of ζ_t.

Assumption 7.1 *In any period t, if there is no strictly positive equilibrium value of ζ_t for buy orders, agents coordinate on the zero-equilibrium and the game ends at the end of the period. If there are multiple equilibrium strategies with strictly positive ζ_t, agents coordinate on the highest such value. This coordination rule is common knowledge.*

We focus on this "most aggressive" equilibrium for the following reason. First, the incentive to trade the currency in period t is an increasing function of the level of the attacks in the future periods. We thus consider the best possible case for a speculative attack. Second, we will present in Section 7.4 policies which prevent the occurrence of an attack. A policy which is effective in preventing attacks under the most favorable assumption will be sufficient to eliminate all attacks.

Arbitrage and the payoff of delay

By Assumption 7.1, the game ends in period $t+1$ if the payoff of an order in that period is not strictly positive. One can show (using an argument in Chamley (2003a)), that in equilibrium, the payoff of delay of the marginal agent with cost $(\lambda_{t-1} + \zeta_t)\bar{c}$ is equal to that of making a final decision in the following period, either to order the asset in period $t+1$ or to stay in the currency until the last period. This *one-step* property is a consequence of Assumption 7.1. (It was found already in the game with no payoff externalities of Chamley and Gale (1994)). The payoff of delay for an agent with cost is therefore equal to

$$w_t(\zeta_t; \lambda_{t-1}, \mu_{t-1}) = \delta \int_{y_t < \bar{X}} U_{t+1}(\lambda_{t-1} + \zeta_t, \mu_t) dF(\theta, \eta; \mu_{t-1}), \qquad (7.5)$$

where $U_{t+1}(\lambda_t, \mu_t)$ is the net expected payoff of ordering the asset in the equilibrium in period $t+1$. (This value is always positive). The functions U_{t+1} and V_{t+1} will be defined by backward induction. Assume now these functions given for the period $t+1$.

[11] Since the incentive to hold is weaker than the incentive to buy, there may be a level of belief such that no speculator buys but asset holders do not sell. The price in period t may convey sufficient information to induce a resumption of new orders in period $t+1$ and eventually a successful attack. While such an equilibrium is theoretically possible, we assume that agents do not coordinate on it.

Since the mass of asset holders is no greater than one, then $\zeta \in [0, 1 - \lambda_{t-1}]$. A necessary condition for ζ to be an equilibrium value in period t is that the payoff of a buy order is at least equal to that of delay:

$$u_t(\zeta; \lambda_{t-1}, \mu_{t-1}) \geq w_t(\zeta; \lambda_{t-1}, \mu_{t-1}). \tag{7.6}$$

There are two cases.

- If there is no $\zeta > 0$ which satisfies (7.6), then the equilibrium value is $\zeta_t = 0$ and by Assumption 7.1, the game ends in period t with the zero-equilibrium. All speculators hold only the currency for all remaining periods.

- If there is a value $\zeta > 0$ such that (7.6) is satisfied, by Assumption 7.1, the equilibrium value ζ_t is defined as

$$\zeta_t = Max\{\zeta \in (0, 1 - \lambda_{t-1}] | u_t(\zeta; \lambda_{t-1}, \mu_{t-1}) \geq w_t(\zeta; \lambda_{t-1}, \mu_{t-1})\}. \tag{7.7}$$

In general, the strategy in period t is a function $\zeta_t = \phi_t(\lambda_{t-1}, \mu_{t-1})$ with

$$\begin{aligned} \phi_t(\lambda_{t-1}, \mu_{t-1}) &= Max\{0, \\ &\quad \{\zeta \in (0, 1 - \lambda_{t-1}] | u_t(\zeta; \lambda_{t-1}, \mu_{t-1}) \geq w_t(\zeta; \lambda_{t-1}, \mu_{t-1})\}\}. \end{aligned} \tag{7.8}$$

When $\zeta_t = \phi_t(\lambda_{t-1}, \mu_{t-1}) > 0$, a speculative attack takes place. It is of one of the following two types: (i) If $\zeta_t = 1 - \lambda_{t-1}$, all agents who have an option place an order. Since there cannot be new orders in period $t + 1$, the attack either succeeds in period t or it fails with all agents selling the foreign asset in period $t + 1$. (ii) If $0 < \zeta_t < 1 - \lambda_{t-1}$, by continuity of u_t and w_t, (7.6) must be an equality, $u_t(\zeta_t; \lambda_{t-1}, \mu_{t-1}) = w_t(\zeta_t; \lambda_{t-1}, \mu_{t-1})$. In such an equilibrium, there is an *arbitrage* between buying in period t at a relatively low price with less information, and delaying until period $t + 1$ to get more information while facing the risk of missing the benefit from a devaluation in period t.

The strategy $\phi_t(\lambda_{t-1}, \mu_{t-1})$ determines the payoff of an order, in equilibrium:

$$U_t(\lambda_{t-1}, \mu_{t-1}) = u_t(\phi_t(\lambda_{t-1}, \mu_{t-1}), \lambda_{t-1}, \mu_{t-1}). \tag{7.9}$$

Value of holding the asset

Suppose that the belief μ is such that some agents place a buy order. In this situation, if some agents find it profitable to place a new order, then the agents who already own the asset and have the same information strictly prefer to hold rather than to sell: since they hold the asset, they do not face probability of rationing when a devaluation takes place. This property is formalized in Lemma 7.1 which can be proven using the same argument as in Chamley (2003b).

Lemma 7.1 *In an equilibrium, if some agents place an order to buy the asset in period t (and $\phi_t(\lambda_{t-1}, \mu_{t-1}) > 0$), then no asset holder sells in period t.*

The value of holding the asset at the beginning of period t depends on whether the speculative attack continues in the period. Using Lemma 7.1, this value satisfies the following recursive equations.

- If $\phi_t(\lambda_{t-1}, \mu_{t-1}) > 0$,

$$\begin{aligned} V_t(\lambda_{t-1}, \mu_{t-1}) &= A \int_{y_t > \bar{X}} dF\left(\theta, \eta; \mu_{t-1}\right) \\ &+ \delta \int_{y_t < \bar{X}} V_{t+1}\left(\lambda_{t-1} + \phi_t, \mu_t\right) dF\left(\theta, \eta; \mu_{t-1}\right). \end{aligned} \tag{7.10}$$

- If $\phi_t(\lambda_{t-1}, \mu_{t-1}) = 0$, by Assumption 7.1 the equilibrium is the zero-equilibrium and

$$V_t(\lambda_{t-1}, \mu_{t-1}) = 0.$$

Backward determination of the equilibrium

In the last period $T + 1$, we have $U_{T+1} \equiv 0$ and $V_{T+1} \equiv 0$. Assume that for $t \leq T$, U_{t+1} and V_{t+1} are given. The payoff of a buy order in period t is determined by (7.4) and that of delay by (7.5). The policy function $\phi_t(\lambda_t, \mu_t)$ is then determined by (7.9). This function determines U_t by (7.9) and V_t by (7.10). The equilibrium is determined for all periods by backward induction.

The previous definition of the policy function generates an equilibrium which is stable in the sense that a small deviation of all currency holders from the equilibrium strategy induces a reaction towards the equilibrium strategy. Consider first the corner solution and assume that $\zeta_t = 1 - \lambda_t > 0$. Ruling out an event with probability zero, $u_t(\zeta_t; \lambda_{t-1}, \mu_{t-1}) > 0 = w_t(\zeta_t; \lambda_{t-1}, \mu_{t-1})$. Assume that a perturbation occurs in the form of a small reduction of ζ_t to $\zeta_t' < \zeta_t$ (or that a small fraction of the currency holders do not speculate). By continuity, it is still true that exercising the option generates a higher payoff, $u_t(\zeta_t'; \lambda_{t-1}, \mu_{t-1})$, than holding the option which has the value $w_t(\zeta_t'; \lambda_{t-1}, \mu_{t-1})$. The optimal response is still to place an order with probability 1.

Suppose now that there is arbitrage with $0 < \phi_t(\lambda_{t-1}, \mu_{t-1}) < 1 - \lambda_t$, and consider the difference between the payoff of an order and that of delay:

$$D(\zeta) = u_t(\zeta; \lambda_{t-1}, \mu_{t-1}) - w_t(\zeta; \lambda_{t-1}, \mu_{t-1}).$$

By definition of ϕ_t, $D(\zeta) < 0$ for all $\zeta \in (\phi_t, 1 - \lambda_{t-1}]$ and its derivative is not equal to 0 at the point $\zeta = \phi_t$. Hence the graph of $u_t(\zeta; \lambda_{t-1}, \mu_{t-1})$ cuts that of $w_t(\zeta; \lambda_{t-1}, \mu_{t-1})$ from above at $\zeta = \zeta_t$: if the total quantity of orders is reduced from the equilibrium value ζ_t to a slightly smaller value ζ_t', the exercise value of the option, u, becomes strictly higher than the holding value w, and the optimal reaction of an individual is to speculate. The reaction is stabilizing in the sense that it operates as an increase in the aggregate quantity of orders[12] towards the equilibrium value ζ_t.

7.2.2 Properties

The general analysis of the model is similar to that in Chamley (2003b). One can show (Property 1 in Chamley (2003b)), that for any fixed period t, the set of values of $(\lambda_{t-1}, \mu_{t-1})$ for which there is an equilibrium with a speculative attack $(\zeta_t > 0)$ is increasing with the number of remaining periods $T-t$. This property shows that a greater number of periods generates more opportunities for information and thus facilitates the coordination of agents in a speculative attack. One can also show as in Property 2 in Chamley (2003b), that in any period t for any $(\lambda_{t-1}, \mu_{t-1})$, the strategy $\phi_t(\lambda_{t-1}, \mu_{t-1})$ of the "most aggressive equilibrium" as specified in (7.7) converges to a stationary solution when the number of remaining periods $T - t$ tends to infinity. The trade-off between immediate investment and delay is now described within the context of an example.

7.3 An example with Gaussian noise

Assume that η has a normal distribution with variance σ_η^2. Let $\nu_t = Log(\mu_t/(1-\mu_t))$ be the log likelihood ratio (LLR) between the two states. The Bayesian equation (7.3) takes the form

$$\nu_t = \nu_{t-1} + \lambda_t \frac{\theta_1 - \theta_0}{\sigma_\eta^2}\left(y_t - \lambda_t \frac{\theta_1 + \theta_0}{2}\right),$$

with $\lambda_t = \lambda_{t-1} + \zeta_t$, and $y_t = \lambda_t \theta + \eta_t$. The expected change of belief from period t to period $t+1$ is measured by

$$E[\nu_{t+1} - \nu_t] = \begin{cases} \frac{(\theta_1-\theta_0)^2}{2\sigma_\eta^2}(\lambda_{t-1} + \zeta_t)^2 \text{ in state } \theta_1, \\ -\frac{(\theta_1-\theta_0)^2}{2\sigma_\eta^2}(\lambda_{t-1} + \zeta_t)^2 \text{ in state } \theta_0. \end{cases} \tag{7.11}$$

We have seen before that the signal to noise ratio in the demand y_t increases with the fraction of active speculators $\lambda_{t-1} + \zeta_t$. This property appears in the previous expression where the absolute value of the expected change of belief increases with the demand by speculators.

A numerical example

The values of the payoff of buying the asset and that of waiting in the first period of a three period game $(T = 2)$ are represented in Figure 7.1[13]. The

[12] The concept of stability remains informal here because there is no room for an out-of-equilibrium ajustment process in real time.

[13] The parameter values are: belief of the high state, $\mu_0 = 0.09$; rate of the devaluation, $A = 0.1$; cost parameter $\bar{c} = 0.015$; standard error of the noise trade, $\sigma_\eta = 0.08\bar{X}$; masses of speculators in the two states, $\theta_0 = 0.6\bar{X}$ and $\theta_1 = 1.2\bar{X}$; discount rate, 0.1 percent. ζ is the fraction of speculators who buy the foreign exchange in the first period. Note that the probability that a devaluation is triggered by the noise traders is negligible.

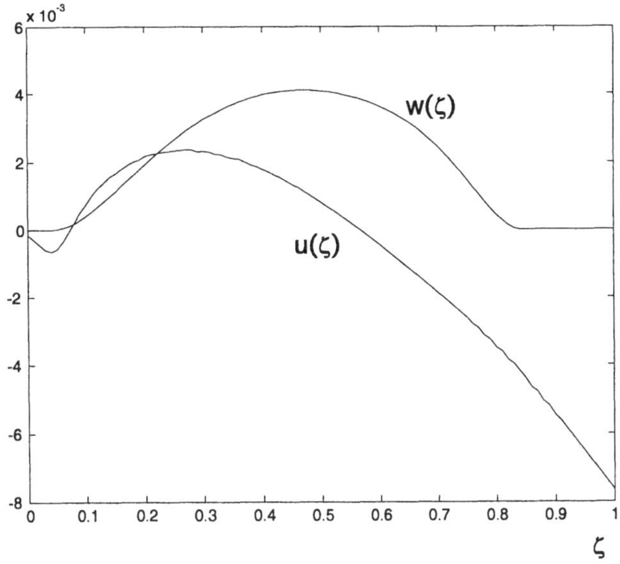

Figure 7.1: Payoff (for the marginal agent) of buying the foreign currency with no delay (u) and payoff delay (w).

payoffs of an order and of delay in the first period are $u(\zeta)$ and $w(\zeta)$ for short[14]. There are three equilibria, but the middle equilibrium value of λ is unstable. The two stable equilibria are the zero-equilibrium and an equilibrium in which a fraction of speculators purchase the asset. (For other values of the parameters, there may be more than one stable equilibrium where the mass of speculation is strictly positive).

When ζ is small, the value of a buy order is a decreasing function of ζ: a higher mass of speculation raises the cost of the marginal agent, but because the mass of endogenous orders is still small, it does not provide much information on the state. The gross payoff of an order remains low and the net payoff of the marginal agent decreases.

When ζ is sufficiently large, the amount of orders conveys an informative signal on the high state if that is the true state. The anticipation that such an information will be provided at the end of the first period and will generate a successful continuation of the attack in the next period raises the value of buying the foreign currency, without delay, above the value of the option of delay. The gain of buying early compensates for the risk of finding out at the end of period one that the bad state is more likely and enduring a loss when the attack stops.

[14] $u(\zeta)$ and $w(\zeta)$ are computed on a grid for $\zeta \in [0,1]$ of width 0.01. The first step is the computation of $U_2(\lambda,\mu)$ and $V_2(\lambda,\mu)$ in the next to last period on a grid of values $(\lambda,\mu) \in [0,1] \times [0,1]$.

7.4 Defense policies

A defense policy is successful if it abolishes the coordination equilibrium with a speculative attack. Two types of policies are considered here: (i) raising the discount rate, (ii) random interventions. For each policy it is assumed that the central bank cannot observe the state θ.

7.4.1 The discount rate

A higher discount rate reduces the discounted benefit of a devaluation which occurs in the future and therefore the payoff of investment in the first period ($u(\zeta)$ in Figure 7.1). The policy reduces the amount of endogenous investment in the first period and therefore the information content of the total investment. For sufficiently high discount rate, the payoff of investment in the first period may be smaller than the option value of delay for any ζ. In this case, the zero-equilibrium is the only equilibrium. A sufficiently high discount rate prevents the occurrence of any speculative attack.

7.4.2 Random intervention

Trades by the central bank which cannot be predicted by speculators have to be random. Assume that the central bank trades a random amount R which is normally distributed $N(0, \sigma_R)$ and set before the opening of the market. Rational speculators know the parameter σ_R but cannot observe R. The random trading by the central bank adds noise and thus reduces the information content of the price in the first period. Numerical simulations show that the smaller information reduces the possibility of coordination between speculators.

In the example of Figure 7.1, the speculative attack is eliminated when σ_η is raised from $0.08\bar{X}$ to $0.105\bar{X}$. This requires a random intervention by the central bank with a standard error equal to $\sqrt{(0.105)^2 - (0.08)^2} = 0.068$. Making the approximation of a range within two standard errors, a successful defense does not necessitate to increase the level of reserves by 20 percent, which is the amount required if the state θ_1 is known. The amount of reserves which is required for a successful defense against speculators under imperfect information is thus smaller than what would be required under perfect information. This is an important implication of the present model for policy. By trading in a non predictable manner, the central bank can prevent speculators from coordinating an attack. In figure 7.2, the parameters are the same as in Figure 7.1, except for σ_η which is raised from $0.08\bar{X}$ to $0.12\bar{X}$.

The central bank may adjust the level of its perturbation in any period t to the fraction of asset holders λ_{t-1} and to the public belief μ_{t-1}. Further work could use the present framework to analyze such a contingent policy and the optimal timing of interventions by the central bank.

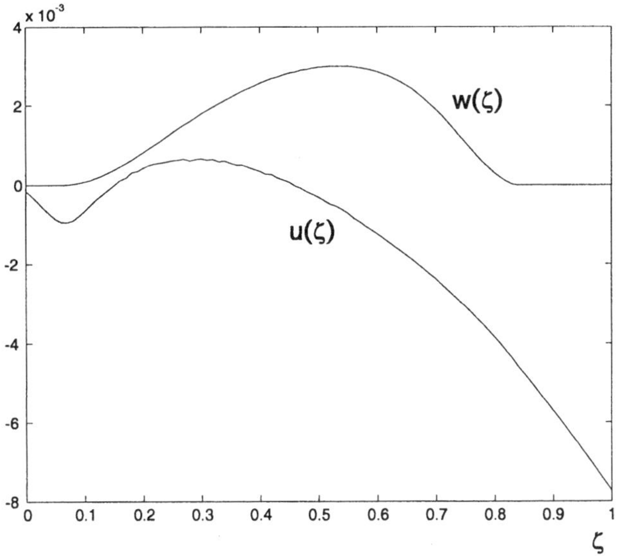

Figure 7.2: Payoffs with additional noise trade.

References

Blume, L., Easley, D. and O'Hara, M. (1994), "Market statistics and technical analysis: The role of volume," *Journal of Finance*, 49, 153-81.

Buiter, W., Corsetti, G. and Pesenti, P., eds. (1998), *Financial markets and European monetary cooperation : the lessons of the 1992-93 exchange rate mechanism crisis*, Cambridge: Cambridge University Press.

Carlsson, H. and van Damme, E.E. (1993a), "Global payoff uncertainty and risk dominance," *Econometrica*, 61, 989-1018.

Carlsson, H. and van Damme, E.E. (1993b), "Equilibrium selection in stag hunt games," in *Frontiers of game theory*, K. Binmore, A. Kirman and A. Tani, eds., Cambridge, MA: MIT Press.

Chamley, C. and Gale, D. (1994), "Information revelation and strategic delay in a model of investment," *Econometrica*, 62, 1065-85.

Chamley, C. (1999), "Coordinating regime switches," *Quarterly Journal of Economics*, 114, 869-905.

Chamley, C. (2003a), "Delays and equilibria with large and small information in social learning,", mimeo, 2000, EHESS-DELTA, *The European Economic Review*, forthcoming.

Chamley, C. (2003b), "Dynamic speculative attacks," *American Economic Review*, 93, 603-621.

Chamley, C. (2003c), *Rational herds*, Cambridge University Press, forthcoming.

Eichengreen, B., Rose, A. and Wyplosz, C. (1996), "Contagious currency crises," NBER Working Paper No. 5681.

Jeanne, O. and Masson, P. (2000), "Currency crisis, sunspots, and Markov-switching regimes," *Journal of International Economics*, 50, 327-350.

Morris, S. and Shin, H.S. (1999). "A theory of the onset of currency attacks," in *Asian Financial Crisis: Causes, Contagion and Consequences*, Agenor, Vines and Weber, eds., Cambridge: Cambridge University Press.

Obstfeld, M. (1996), "Models of currency crises with self-fulfilling features," *European Economic Review*, 40, 1037-1047.

Rodionova, I. and Surti, J. (1999), "Speculative currency crises in a dynamic setting", mimeo, Boston University.

Notes on the editors

Sandrine Lardic is Senior Lecturer in Economics at the University of Paris X (France) and is a member of the MODEM-CNRS research group. She also is consultant for Sinopia Asset Management. She carries on research concerning econometrics in the fields of financial markets and macroeconomics. She is the author of books and papers on theoretical and applied econometrics.

Valérie Mignon is Professor of Economics at the University of Paris X (France) and is a member of the THEMA-CNRS research group. She conducts research on theoretical econometric methods and applied econometrics in the fields of financial markets and macroeconomics. Her research centers concern mainly long-memory processes and nonlinear econometrics. She is the author of books and papers on theoretical and applied econometrics.

Index